CD.ATL.538

PreTest®

Histology and
Cell Biology

DISCARD

Notice

Medicine is an ever-changing science. As new research and clinical experience broaden our knowledge, changes in treatment and drug therapy are required. The author and the publisher of this work have checked with sources believed to be reliable in their efforts to provide information that is complete and generally in accord with the standards accepted at the time of publication. However, in view of the possibility of human error or changes in medical sciences, neither the editor nor the publisher nor any other party who has been involved in the preparation or publication of this work warrants that the information contained herein is in every respect accurate or complete, and they are not responsible for any errors or omissions or for the results obtained from use of such information. Readers are encouraged to confirm the information contained herein with other sources. For example and in particular, readers are advised to check the product information sheet included in the package of each drug they plan to administer to be certain that the information contained in this book is accurate and that changes have not been made in the recommended dose or in the contraindications for administration. This recommendation is of particular importance in connection with new or infrequently used drugs.

Histology and Cell Biology

PreTest®
Self-Assessment
and Review

Second Edition

Robert M. Klein, Ph.D.
Professor
Department of Anatomy and Cell Biology
University of Kansas Medical Center
Kansas City, Kansas

 McGraw-Hill
Health Professions Division
PreTest® Series

New York   St. Louis   San Francisco   Auckland
Bogotá   Caracas   Lisbon   London   Madrid
Mexico City   Milan   Montreal   New Delhi
San Juan   Singapore   Sydney   Tokyo   Toronto

McGraw-Hill

A Division of The **McGraw·Hill** *Companies*

Histology and Cell Biology: PreTest® Self-Assessment and Review, Second Edition
Copyright © 1996 1993 by The McGraw-Hill Companies, Inc. All rights reserved.
Printed in the United States of America. Except as permitted under the Copyright Act
of 1976, no part of this publication may be reproduced or distributed in any form or
by any means, or stored in a data base or retrieval system, without the prior written
permission of the publisher.

1 2 3 4 5 6 7 8 9 0 DOCDOC 9 8 7 6 5

ISBN 0-07-052081-X

The editors were Gail Gavert and Bruce MacGregor.
The production supervisor was Gyl A. Favours.
This book was set in Times Roman by Compset, Inc.
R.R. Donnelley & Sons was printer and binder.

Library of Congress Cataloging-in-Publication Data
Klein, Robert M. (Robert Melvin)
  Histology and cell biology : PreTest self-assessment and review/
Robert M. Klein, — 2nd ed.
    p. cm.
  Includes bibliographical references.
  ISBN 0-07-052081-X
  1. Histology—Examinations, questions, etc. 2. Cytology—
Examinations, questions, etc. I. Title.
  [DNLM: 1. Cytology—examination questions. 2. Histology—
examination questions.   QS 18.2 K64h   1996]
QM554.H58 1996
611'.0076—dc20
DNLM/DLC
for Library of Congress   95-5739
                                                                  CIP

Contents

Contents

To my wife, Beth, and our children, Melanie, Jeffrey, and David,
for their support and patience during the writing of this text,
and to my parents, Nettie and David,
for their emphasis on the pursuit of knowledge

Preface

In this second edition of *Histology and Cell Biology: PreTest® Self-Assessment and Review*, a significant number of changes and improvements have been made. An additional cell biology section has been developed to cover the topic of intracellular trafficking. This new chapter reviews processes such as endocytosis, exocytosis, and signal transduction as well as the signals that permit various molecules to gain access to specific subcellular compartments. While the second edition of this text continues to focus on cell biology and the application of principles of cell biology to the study of tissues and organ systems, there have also been important revisions of the text in the chapters dealing with histology. Coverage of this area has been extensively expanded to provide more light and electron micrographs. Topics and references have been updated to include important new knowledge in cell and tissue biology. There is also a greater focus on clinically related questions and problems. The number of items per group of matching questions has been reduced in accordance with the format used on USMLE Step 1.

The author expresses his gratitude to his colleagues who have greatly assisted him by providing light and electron micrographs as well as constructive criticism and revisions of the line drawings. He also acknowledges Karen Chinn for her painstaking care in the preparation of the line drawings. The author is indebted to the students of the University of Kansas School of Medicine, past and present, who have challenged him to continuously improve his skills as an educator.

Introduction

Each *PreTest® Self-Assessment and Review* allows medical students to comprehensively and conveniently assess and review their knowledge of a particular basic science, in this instance Histology and Cell Biology. The 500 questions parallel the format and degree of difficulty of the questions found in the United States Medical Licensing Examination (USMLE) Step 1. Practicing physicians who want to hone their skills before USMLE Step 3 or recertification may find this to be a good beginning in their review process.

While the content of traditional histology courses is reviewed in this book, the questions reflect curricular changes that have occurred in this discipline in medical schools across the United States during the last decade. Histology courses have been modified to introduce cell biology and the application of principles of cell biology to the study of tissues and organs. The first four sections of this book cover the cell biology of the plasma and intracellular membranes, cytoplasm, intracellular trafficking, and the nucleus. The basic tissues of the body—including epithelial, connective, specialized connective (bone and cartilage), muscle, and neural tissues—are covered in subsequent sections. The cell biology of the blood, bone marrow, and cardiovascular system is discussed in a separate chapter. Following the review of tissues, questions are provided in separate sections for each of the organ systems: lymphoid, respiratory, integumentary, gastrointestinal (including gastrointestinal glands), endocrine, urinary, male and female reproductive, and the eye and ear. For each system, light and electron micrographs have been included to facilitate the review of appropriate morphological concepts. In many areas, questions provide an integrated approach to coverage of a tissue or organ system which encompasses physiology, pathology, and biochemistry as well as the cell biology and microscopic structure of these systems. Where appropriate, clinically related problems and questions are included in each chapter.

Each question is accompanied by an answer, a detailed explanation, and a specific page reference to an appropriate textbook or journal article. A bibliography listing sources can be found following the last chapter of this text.

An effective way to use this PreTest is to allow yourself one minute to answer each question in a given chapter. As you proceed, indicate your answer beside each question. By following this suggestion, you approximate the time limits imposed by the Step examination.

After you finish going through the questions in the section, spend as much time as you need verifying your answers and carefully reading the explanations provided. Pay special attention to the explanations for the questions you answered incorrectly—but read *every* explanation. The authors of this material have designed the explanations to reinforce and supplement the information tested by the questions. If you feel you need further information about the material covered, consult and study the references indicated.

Cell Biology: Membranes

DIRECTIONS: Each question below contains five suggested responses. Select the **one best** response to each question.

1. The basic structure of biologic membranes

(A) is visualized as a bilaminar structure with transmission electron microscopy

(B) differs ultrastructurally for internal and external membranes of the cell

(C) possesses a thickness of 1 to 2 mm

(D) is a lipid bilayer that serves as a barrier to water-soluble molecules

(E) is best described as lipids dispersed within a protein bilayer

2. Which of the following is true of the carbohydrate portion of the membrane?

(A) It is found primarily in the form of free saccharide groups

(B) It is located on the cytosolic side of the plasma membrane

(C) It forms the glycocalyx consisting of glycosaminoglycans, glycolipids, and glycoproteins

(D) It has a symmetric distribution

(E) It is found on the outside (cytosolic side) of organellar membranes, such as that of the mitochondrion

3. The face labeled by asterisks in the freeze-fracture preparation shown below may be characterized as

From Fawcett DW: *The Cell,* 2/e, WB Saunders, 1981, with permission.
Courtesy of Dr. G. Raviola.

(A) containing primarily glycoproteins and glycolipids
(B) facing away from the cytoplasm
(C) in direct contact with the cytoplasm
(D) in direct contact with the external surface
(E) generally possessing a paucity of intramembranous particles

4. Which of the following statements correctly characterizes membrane protein?

(A) Content is constant in cells with different functions

(B) Polypeptide chains can extend across the lipid bilayer once or multiple times

(C) Integral membrane proteins can be removed entirely from the membrane by mild extraction methods

(D) Membrane protein includes integral proteins associated with the extracellular surface

(E) Membrane protein includes peripheral proteins bound to phospholipids in the membrane bilayer

5. Glycophorin is a single-pass, transmembrane glycoprotein found in the erythrocyte (RBC). Which of the following is an expected characteristic of this protein?

(A) It possesses a hydrophilic portion that spans the lipid bilayer

(B) Hydrolysis of carbohydrate will not affect the glycophorin band on an SDS polyacrylamide gel

(C) It possesses oligosaccharides on the cytosolic side of the membrane

(D) The polypeptide chain crosses the lipid bilayer in an α-helix conformation

(E) It may be isolated from the RBC membrane with mild extraction conditions such as altered pH or ionic strength

DIRECTIONS: Each numbered question or incomplete statement below is NEGATIVELY phrased. Select the **one best** lettered response.

6. All the following statements are true of the fluid-mosaic membrane model of the cell membrane EXCEPT

(A) this model visualizes the plasma membrane as a sea of lipid that takes on the form of a phospholipid layer
(B) the functional elements of this membrane model are the proteins found in the lipid bilayer
(C) lipids form the foundation of the bilayer
(D) proteins are arranged in a continuous layer in the cell membrane
(E) cholesterol content is a major factor in the determination of membrane fluidity

7. All the following contribute to membrane fluidity EXCEPT

(A) rotational movement (diffusion) of proteins and lipids in the membrane
(B) lateral movement (diffusion) of proteins and phospholipids
(C) transbilayer movement of phospholipids in the endoplasmic reticulum
(D) increased cholesterol content of the plasma membrane
(E) binding of ligands to receptors resulting in patching and capping phenomena

8. All the following contribute to the polyanionic charge on the extracellular surface EXCEPT

(A) polar, hydrophilic ends of the phospholipid bilayer
(B) glycosaminoglycans adsorbed on the extracellular surface
(C) sialic acid–rich oligosaccharides
(D) transmembrane proteoglycans
(E) the sulfhydryl groups of transmembrane glycoproteins

9. With regard to the asymmetry of cell membranes, all the following statements are true EXCEPT

(A) cholesterol is asymmetrically distributed within the bilayer
(B) asymmetry is established during membrane synthesis (i.e., in the endoplasmic reticulum)
(C) phospholipids follow an asymmetric distribution within the bilayer
(D) the N-terminus of transmembrane protein extends from the external surface of the plasma membrane
(E) peripheral proteins are restricted to the internal (cytosolic) surface of the plasma membrane

10. Considerable information about membrane function has been obtained from red blood cell membranes. All the following would be expected to limit or reduce lateral diffusion of membrane proteins within the plane of the lipid bilayer EXCEPT

(A) binding to the spectrin membrane skeleton via ankyrin
(B) binding to fibronectin
(C) binding to laminin
(D) binding to the actin-based cytoskeleton by proteins such as talin
(E) binding of an antibody to a cell surface antigen

11. Receptors such as the muscarinic, cholinergic receptor rhodopsin and the β-adrenergic receptor are members of the family of G protein–linked receptors. These receptors are multiple-pass transmembrane proteins and are characterized by all the following EXCEPT

(A) the formation of a gated channel when the ligand binds the receptor
(B) an arrangement of hydrophilic membrane-spanning segments
(C) an intracellular carboxyl terminal
(D) extracellular N-linked glycosylation sites
(E) extracellular and intracellular connecting loops

12. All the following statements are true concerning plasma membrane domains EXCEPT

(A) membrane protein of the apical domain does not normally intermix with the basolateral domain
(B) membrane domains can be demonstrated by freeze-fracture techniques
(C) the lipid and protein contents of the apical and basolateral domains are virtually identical
(D) lectin-ferritin conjugates can be used to demonstrate membrane domains
(E) disruption of tight junctions can cause intermixing of membrane protein from one membrane domain to another

Cell Biology: Membranes

Answers

1.   The answer is D. *(Alberts, 3/e, pp 475–479. Goodman, pp 25–28. Junqueira, 8/e, pp 20–25. Ross, 3/e, pp 20–21. Widnell, pp 27–30.)* Cell membranes range in thickness between 7 and 10 nm (1 mm = 10^{-6} m, 1 nm = 10^{-9} m; the diameter of a red blood cell is 7 mm). A cell membrane surrounds all eukaryotic cells and subserves a number of essential functions. The plasma membrane forms a boundary to the external environment and contains a large variety of receptors that function as ligands for hormones, growth factors, cytokines, and other extracellular factors. It is a selective barrier that controls the entry and exit of substances to and from the cell and is involved in both energy-dependent (active) and energy-independent (passive) transport and the generation of ionic gradients between the inside and outside of the cell. The plasma membrane and the internal membranes of the cell, such as those surrounding organelles like the nucleus and mitocondria, are similar in their morphological appearance in the electron microscope. The organellar membranes are usually thinner and differ in overall biochemical composition from the plasma membrane.

The plasma membrane is visualized as a trilaminar structure in electron micrographs in which the lipid bilayer is stained with osmium. The trilaminar membrane is the classic form of the membrane whether one is viewing the plasma membrane or the membrane surrounding the intracellular organelles. The two dense lines with a light-line in between gives the appearance of a railroad track. The membrane actually consists of a bilayer of phospholipids with the nonpolar, hydrophobic layer in the central portion of the membrane. The outer portion of the membrane is, therefore, hydrophilic as the polar regions of the phospholipids are in contact with aqueous components at the intra- or extracellular surfaces of the membrane. This lipid bilayer is therefore amphiphatic (containing both hydrophobic and hydrophilic portions) and is responsible for the impermeability to water-soluble molecules that is exhibited by cellular membranes. Carbohydrates are found in the membrane in the form of glycolipids and glycoproteins. Proteins are generally dispersed within the lipid bilayer.

2.   The answer is C. *(Alberts, 3/e, pp 483–485, 502. Goodman, pp 28, 30. Junqueira, 8/e, pp 22–23, 66. Ross, 3/e, pp 20–23. Widnell, pp 33–34.)* The carbohydrate of biologic membranes is found in the form of glycoproteins and glycolipids rather than as free saccharide groups. There is an asymmetric distribution of carbohydrate within biologic membranes. Carbohydrate is found specifically on the noncytoplasmic surface (leaflet) of the membrane. In the plasma membrane, therefore, sugars are found predominantly on the outside of the cell and in organellar membranes on the luminal surface. The glycocalyx ("chalice of sugar"), or cell coat, is the fuzzy region on the extracellular side of the membrane and has a high carbohydrate content. This zone is visible with the electron microscope and by using special stains (e.g., ruthenium red) or lectins, carbohydrate-binding proteins that detect the presence of carbohydrates on the cell surface. Lectins may be bound to fluorescent or other markers to allow visualization with specialized microscopic techniques. The glycocalyx is prominent on epithelial cells such as those lining the lumen of the small intestine, but is not limited to this cell type. The carbohydrates found in the form of glycoproteins and glycolipids contribute to the polyanionic charge on the extracellular surface.

3. The answer is B. *(Alberts, 3/e, pp 153, 494–495. Goodman, pp 29–31. Junqueira, 8/e, p. 24. Ross, 3/e, pp 23–24. Widnell, pp 323–327.)* Freeze fracture is a preparative procedure in which the tissue is rapidly frozen and fractured with a knife. The fracture plane occurs through the hydrophobic central plane of membranes, which is the plane of least resistance to the cleavage force. The two faces are essentially the two interior faces of the membrane. They are described as the extracellular face (E face) and the protoplasmic face (P face). The P cytoplasm is the backing for the face, which in general contains numerous intramembranous particles. The E face is backed by the extracellular space and in general contains a paucity of intramembranous particles (see upper part of figure) compared with the P face (labeled with asterisks). Glycolipids and glycoproteins compose the glycocalyx that covers the cell membrane on its exterior surface.

4. The answer is B. *(Alberts, 3/e, pp 485–486, 488–489. Goodman, pp 28– 30. Junqueira, 8/e, pp 22–25. Ross, 3/e, pp 20–23. Widnell, pp 28–30.)* Membrane protein is the variable portion of the cell membrane, while lipid content remains fairly constant. The result is variation in the protein/lipid ratio. The average protein mass in membranes is about one-half but varies from one-fourth to three-fourths, and the lipid/protein ratio varies, depending upon cell function. Polypeptide chains that constitute transmembrane proteins may be single-pass or multiple-pass molecules as they cross the membrane from the cytosolic to the extracellular side. Some proteins may be removed from the lipid bilayer through mild extraction techniques that use alteration of surface charge or pH. These proteins are the peripheral (extrinsic) membrane proteins associated with the extracellular surface. In contrast, the second type of membrane protein is the integral (intrinsic) membrane proteins, which are bound tightly to components of the phospholipid bilayer. Integral membrane proteins require detergent treatment for removal from the membrane. Detergents are amphipathic molecules; that is, they have both hydrophobic and hydrophilic portions, which facilitates membrane disruption.

5. The answer is D. *(Alberts, 3/e, pp 493–494. Goodman, pp 29–30.)* Glycophorin is a transmembrane glycoprotein found in the red blood cell. As a transmembrane protein it traverses the membrane and crosses the lipid bilayer in a single-pass α-helix conformation. The hydrophobic portion spans the lipid bilayer and the hydrophilic carboxyl end is exposed to the cytosol, while the hydrophilic amino end is exposed to the extracellular surface. The oligosaccharides are found on the hydrophilic amino terminus where the negative surface charge is generated. They are degraded by carbohydrate hydrolysis. Harsh detergent treatment is required for isolation of transmembrane proteins such as glycophorin.

6. The answer is D. *(Alberts, 3/e, pp 477–482. Goodman, pp 25–28. Junqueira, 8/e, pp 22–25. Ross, 3/e, p 21. Widnell, pp 27–34.)* In the fluid-mosaic model originally proposed by Singer and Nicholson, the plasma membrane is viewed as a sea of lipid in the form of a phospholipid bilayer. Proteins do not form a regular, continuous layer within the membrane but float in the lipid as "icebergs" that drift relatively independently. These proteins subserve the functions of enzymes, receptors, or transport, while lipids form the foundation of the membrane. Cholesterol reduces the fluidity of the lipid bilayer by interacting with the portion of the hydrophobic chains that are closest to the hydrophilic, polar head groups and stiffening this region of the membrane.

7. The answer is D. *(Alberts, 3/e, pp 480–482, 498–499. Goodman, pp 32–33. Junqueira, 8/e, pp 22–25. Widnell, pp 34–36.)* Rotational and lateral movements of both proteins and lipids contribute to membrane fluidity. Phospholipids are capable of lateral diffusion, rapid rotation around their long axis, and flexion of their hydrocarbon (fatty acyl) tails. They undergo transbilayer movement, known as "flip-flop," between bilayers in the endoplasmic reticulum; however, this is a very rare occurrence in the plasma membrane. Proteins also undergo rotational and lateral diffusion. For example, when a divalent or multivalent ligand binds to membrane receptors present as intrinsic membrane proteins, initially there is a homogeneous pattern to the binding. Subsequently, the ligand-receptor complexes undergo patching and eventually capping on the cell surface. The processes of patching and capping are excellent examples of the lateral diffusion of protein molecules within the membrane. Fluorescence recovery after photobleaching is used to measure the lateral mobility of membrane protein. Bleaching of fluorescence is induced with a laser and the time required for bleached and unbleached molecules to mix is known as the *diffusion coefficient.* For rhodopsin, the photoreceptive protein of rod cells in the retina, the diffusion coefficient is the highest of any known membrane protein (5×10^{-9}cm^2/s). Lipids also undergo lateral diffusion.

Other factors reduce membrane fluidity. Cholesterol increases membrane rigidity by interacting with the hydrophobic regions near the polar head groups and stiffening this region of the membrane. Association or binding of integral membrane proteins with cytoskeletal elements on the interior and peripheral membrane proteins on the extracellular surface may serve to limit membrane mobility and fluidity.

8. The answer is E. *(Alberts, 3/e, pp 483–484, 502–503. Widnell, pp 27–34.)* The polyanionic surface charge occurs by the presence of a combination of the polar ends of the phospholipid bilayer; glycosaminoglycans, which may be secreted and subsequently adsorbed to the surface; oligosaccharides associated with glycoproteins and glycolipids; and transmembrane proteoglycans that have negatively charged oligosaccharides attached to them on their carboxyl end. The sulfhydryl groups of transmembrane glycoproteins are found on the cytosolic surface and are characterized by a reducing environment in which sulfhydryl groups remain; i.e., disulfide bonds are not formed in the cytosolic domains.

9. The answer is A. *(Alberts, 3/e, pp 482–485. Goodman, pp 30–33. Junqueira, 8/e, pp 22–24. Widnell, pp 27–34.)* Cholesterol is different from proteins and phospholipids, which are asymmetrically distributed within the bilayer. Cholesterol is found on both sides of the bilayer. The small polar head group structure of cholesterol allows it to flip-flop from leaflet to leaflet and respond to changes in shape. In contrast to cholesterol, most proteins and phospholipids are capable of only rare flip-flop. For example, transbilayer movement of phospholipid is mostly limited to the endoplasmic reticulum. Peripheral membrane proteins such as spectrin are associated specifically with the cytosolic surface of the cell membrane. Phospholipids are also specifically associated with one of the leaflets. Carbohydrates are associated with the N-terminal of transmembrane protein, which extends from the extracellular surface. The C-terminus of transmembrane protein extends into the cytoplasm. Asymmetry of the lipid bilayer is established during membrane synthesis in the endoplasmic reticulum. One of the best examples is the presence of protein kinase C in the cytoplasmic leaflet of the phospholipid bilayer.

10. The answer is E. (*Alberts, 3/e, pp 491–493. Goodman, pp 29, 84. Stevens, p 19. Widnell, pp 82–83.*) Spectrin is a cytoskeletal protein noncovalently bound to the cytosolic side of the membrane in red blood cells. It is found in high concentrations in red blood cell membranes. Actin readily forms complexes with spectrin, which may be considered as an intermediate filament protein. Spectrin is bound to another protein, ankyrin, which attaches to transmembrane band 3 and limits lateral diffusion. Binding of cell membranes to extracellular matrix molecules such as fibronectin and laminin through transmembrane proteins such as integrins (i.e., fibronectin and laminin receptors) also reduces lateral diffusion by tethering the membrane to the extracellular matrix. Talin is a protein that anchors actin, a major cytoskeletal element, to transmembrane proteins such as the integrins. The binding of an antibody to a cell surface antigen should result in patching and eventual capping on the extracellular surface as a result of ligand-receptor binding. This process involves an increase in lateral mobility of proteins in the lipid bilayer. (There will be questions dealing with the cytoskeletal proteins in the cytoplasm section of cell biology and further information about integrins in the epithelial and connective tissue sections of this review book.)

11. The answer is B. (*Alberts, 3/e, pp 486–488, 734–735. Goodman, pp 28–30. Widnell, pp 82–83, 228.*) There is a remarkable homology between the cell surface receptors linked to the G proteins. Included in this group are the muscarinic, cholinergic receptors, rhodopsin, and the β-adrenergic receptor. The receptors are all multipass transmembrane proteins consisting specifically of seven hydrophobic spanning segments of the single polypeptide chain. Between the segments the polypeptide chain loops on both the extracellular and intracellular sides of the membrane. All of these transmembrane proteins demonstrate a carboxyl terminus on the cytosolic side and extracellular *N*-linked glycosylation sites on the extracellular surface. Gated channels are associated with the binding of ligand to receptor as occurs at the acetylcholine receptor.

12. The answer is C. (*Alberts, 3/e, pp 500–503. Goodman, pp 30–32.*) The internal and external membranes of the cell are not homogeneous. For example, in the case of the cell membrane, confinement of proteins and lipids to specific apical and basolateral domains occurs. One of the best examples is the epithelial cell, the polarity of which will be discussed in a subsequent section on the epithelium. Freeze fracture splits the membrane along the central hydrophobic plane of the phospholipid bilayer and demonstrates differences in the presence of intramembranous particles. It has generally been shown that tight junctions of the junctional complex between cells serve to separate apical and basolateral domains, which differ in both lipid and protein content. Lectins bind to carbohydrates, which are cell surface molecules essential to normal cell-cell and cell-matrix interactions. In the case of an epithelial cell, an apical domain in close proximity to the lumen and the basolateral membrane adjacent to another cell in the epithelium are dissimilar in composition.

Cell Biology: Cytoplasm

DIRECTIONS: Each question below contains five suggested responses. Select the **one best** response to each question.

13. The presence of vimentin in immunocytochemical analysis of a metastatic tumor would suggest an origin from

(A) epithelium
(B) neurons
(C) glia
(D) mesenchyme
(E) endoderm

14. Ribosomes are correctly characterized by which of the following statements?

(A) They consist entirely of rRNA
(B) They are highly active when observed in the cytoplasm as single units
(C) They differ in response to antibiotics depending on whether they are eukaryotic or prokaryotic in origin
(D) They move along tRNAs and read the genetic message that has been transcribed from DNA
(E) They are composed of two equal subunits

15. Which of the following statements best describes the functional characteristics of lysosomes?

(A) They only function within the phagocytic pathway of the cell
(B) They contain enzymes that lack macromolecular specificity
(C) They function only within the intracellular compartment
(D) They function at alkaline pH
(E) Some cytosolic proteins contain signals directing them to lysosomes

16. The stability and arrangement of actin filaments as well as their properties and functions depend on

(A) the structure of the actin filaments
(B) microtubules
(C) intermediate filament proteins
(D) actin-binding proteins
(E) motor molecules, such as kinesin

17. The structures marked with the arrows are correctly characterized by which of the following statements?

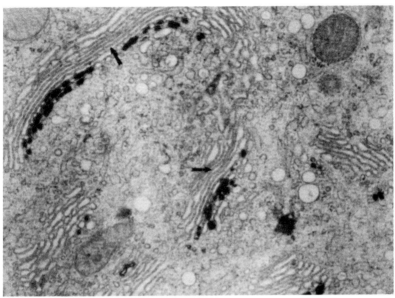

From Fawcett DW: *The Cell*, 2/e. WB Saunders, with permission. Courtesy of Dr. Daniel Friend.

(A) They present an entry face associated with granule formation
(B) They present an exit face associated with transport vesicles
(C) They receive proteins but not lipids
(D) They are biochemically compartmentalized
(E) Topologic organization of the stacks is unrelated to the function of specific enzymes

18. Which of the following is true of microtubules?

(A) Taxol binds to tubulin molecules and prevents polymerization

(B) Colchicine binds to microtubules and stabilizes them

(C) Taxol and colchicine both function as antimitotic drugs

(D) Microtubules are usually found in the cytoplasm as large bundles and extensive, cross-linked networks

(E) Microtubules are highly stable structures that turn over very slowly

19. The primary function of intermediate filaments is to

(A) generate movement

(B) provide mechanical stability

(C) carry out nucleation of microtubules

(D) stabilize microtubules against disassembly

(E) transport organelles within the cell

DIRECTIONS: Each numbered question or incomplete statement below is NEGATIVELY phrased. Select the **one best** lettered response.

20. All the following statements are true of the vesicular structure marked throughout the cytoplasm in the electron micrograph below EXCEPT

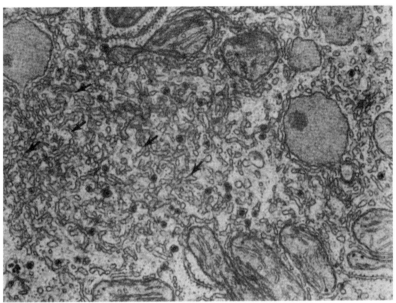

From Fawcett DW: *The Cell,* 2/e. WB Saunders, 1981, with permission. Courtesy of Dr. Robert Bolender.

(A) it can be separated by differential centrifugation from other cytoplasmic components
(B) it is involved in the regulation of muscle contraction
(C) it is the primary site of protein synthesis in the cell
(D) it is involved in lipid synthesis
(E) it is involved in detoxification processes in the liver

21. In regard to the mechanisms by which mitochondria convert energy to the metabolic needs of the cell, all the following are true statements EXCEPT

(A) mitochondria trap chemical energy and store it in the form of ATP
(B) the respiratory chain pumps protons into the mitochondrial matrix
(C) mitochondria maintain an electrochemical proton gradient that represents stored energy
(D) ATP synthetase phosphorylates ADP in response to the flow of protons from the matrix to the inner mitochondrial membrane
(E) there are major enzyme complexes within the respiratory chain that transfer electrons from NADH to O_2

22. Treadmilling, a dynamic feature of some cytoskeletal elements, is characterized by all the following statements EXCEPT

(A) it occurs in vitro for both actin filaments and microtubules
(B) it is dependent on the energy provided by ATP hydrolysis
(C) it depends on identical rates of association and dissociation at the plus and minus ends of the polymer, respectively
(D) it is also known as *nucleation*
(E) it does not occur in the elongation of intermediate filaments

23. All the following statements are true of the endoplasmic reticulum (ER) EXCEPT

(A) it is involved in the synthesis of only exportable protein
(B) it is found in all eukaryotic cells
(C) it is involved in lipid synthesis
(D) it varies in density depending upon the functional activity of the cell
(E) it is continuous with the nuclear envelope

24. In Zellweger syndrome, tissue shows the presence of empty peroxisomes. In these patients, findings one might expect include all the following EXCEPT

(A) liver abnormalities
(B) an inability to detoxify alcohol
(C) renal abnormalities
(D) the absence of a peroxisomal signal peptide or membrane receptor
(E) decreased lysosomal activity

25. Peroxisomes (microbodies) *differ* from mitochondria in all the following ways EXCEPT

(A) they are surrounded by only a single membrane
(B) they do not contain DNA
(C) they do not contain ribosomes
(D) they import all of their proteins and membrane lipids
(E) new organelles form by growth and fission of existing organelles

26. Cytochalasins inhibit actin assembly. They would be expected to interfere with all the following functions EXCEPT

(A) phagocytosis
(B) cytokinesis
(C) development of lamellipodia at the leading edge of a migrating cell
(D) separation of the chromosomes in anaphase of the cell cycle
(E) cell motility

27. Chloroquine is a weak base that neutralizes acidic organelles. Which of the following would be LEAST likely to be affected by chloroquine treatment?

(A) Phagosomes
(B) Lysosomes
(C) Secretory vesicles
(D) Some compartments of the Golgi
(E) Membrane-bound ribosomes

DIRECTIONS: Each group of questions below consists of lettered headings followed by a set of numbered items. For each numbered item select the **one** lettered heading with which it is **most** closely associated. Each lettered heading may be used **once, more than once, or not at all.**

Questions 28–30

Match each description below with the appropriate label in the figure.

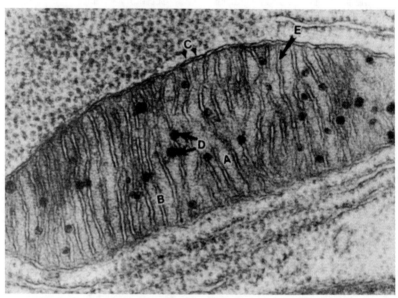

From Fawcett DW: *The Cell,* 2/e. WB Saunders, with permission.

28. Contains the mitochondrial genome

29. Contains the enzymes responsible for the oxidative reactions of the electron transport chain

30. Contains the Krebs cycle enzymes

Questions 31–33

Match each description with the appropriate structure.

- (A) Microfilaments (thin filaments)
- (B) Intermediate filaments
- (C) Microtubules
- (D) Thick filaments
- (E) Spectrin heterodimers

31. Principal structural element of the axoneme

32. An ATPase activated by actin

33. Mediators of kinesin-dependent vesicular transport

Questions 34–35

Match each description with the appropriate substance.

- (A) Clathrin
- (B) Spectrin
- (C) Ankyrin
- (D) Actin
- (E) Vimentin

34. Forms coating of coated pits and vesicles

35. Is located both in the membrane skeleton of red blood cells and in thin microfilaments

Questions 36–38

Match each description with the correct Golgi compartment.

- (A) *Trans*-compartment
- (B) Medial compartment
- (C) *Cis*-compartment
- (D) *Trans*-Golgi network
- (E) Transitional elements

36. Structure that transfers proteins and lipids to the Golgi from the endoplasmic reticulum

37. Part of Golgi that receives proteins and lipids from the endoplasmic reticulum

38. The dispatcher or sorting station for proteins

Cell Biology: Cytoplasm

Answers

13. The answer is D. *(Alberts, 3/e, pp 798–800. Goodman, pp 86–88.)* The heterogeneity of intermediate filament protein subunits is relatively specific for cells derived from the three embryonic germ layers. Antibodies to intermediate filament proteins have been used by pathologists to determine the origin of tumors. Cytokeratins (also known as *keratins*) are specific for epithelial cells; desmin is found in striated and most smooth muscle, except for smooth muscle present in the vasculature; vimentin is specific for mesenchymal cells such as fibroblasts, macrophages, endothelial cells, and smooth muscle of the vasculature; and glial fibrillary acidic protein (GFAP) is specific for astrocytes, not microglia. Neurofilament proteins (NF_L, NF_M, NF_H) are found in neurons and appear to be involved in the anchorage of the proteins that form ion channels in addition to the normal structural role of intermediate filaments. In Alzheimer's senile dementia syndrome, extensive plaques of neurofilament proteins occur.

14. The answer is C. *(Alberts, 3/e, pp 231–234, 236, 240–241. Goodman, pp 113–114. Junqueira, 8/e, pp 29–31.)* Ribosomes are spheroidal, cytoplasmic particles responsible for the synthesis of proteins. They are composed of both protein and RNA (predominantly rRNA, but also mRNA and tRNA). When found as single units in the cytoplasm, they are relatively inactive; they are active when they form polyribosomes with linkage by mRNA. The ribosomes move along mRNAs and read the genetic message transcribed from DNA. The tRNAs bring specific amino acids to the polyribosome complex. As the message is translated by the ribosome, the appropriate amino acid encoded for by the codon read by the ribosome is carried to the region and binds to the ribosome. The peptide is synthesized as a series of amino acids added to the previously synthesized portion of the peptide. Ribosomes are composed of two unequal subunits. The size of the subunits and their response to antibiotics differ between bacterial and cellular ribosomes. It is interesting to note that mitochondrial ribosomes resemble prokaryotic (bacterial) ribosomes in their response to antibiotics.

15. The answer is E. *(Alberts, 3/e, pp 610–611, 614. Junqueira, 8/e, pp 33–37. Ross, 3/e, pp 31–32, 35–36. Stevens, pp 16–18.)* Lysosomes are organelles found in all eukaryotic cells, but they are found in higher numbers in cells that are very active in phagocytosis, such as macrophages and polymorphonuclear leukocytes (PMNs). They stain for acid phosphatase and have a high content of hydrolytic enzymes, but show a dramatic variation in morphology depending upon the content of the material undergoing degradation within the lysosome and the extent of the degradative process. Lysosomes may obtain material for degradation from the endocytic, autophagic, or phagocytic pathways. In this way lysosomes can degrade materials in the intracellular and extracellular pathways, including cell organelles. Some cytosolic proteins contain a signal sequence (KFERQ—lysine, phenylalanine, glutamate, arginine, and glutamine). This sequence results in direct transport to lysosomes for degradation.the lysosomes are membrane-limited, which prevents their acidic enzymes from degrading the cytoplasm. There are a multitude of enzymes found in lysosomes (proteases, glycosidases, lipases, phosphatases, sulfatases, phospholipases, lipases, and nucleases), and these enzymes are specific for many classes of macromolecules (e.g., proteins, lipids, nucleic acids and other molecules). The products of lysosomal breakdown are transported to the cytosol, where they may be reutilized or excreted. Lysosomes may release their enzymes to the extracellular milieu under normal conditions that require breakdown of the extracellular matrix. Osteoclasts, which will be discussed in the section on the specialized connective tissues, degrade the matrix of bone by creating an acidic extracellular milieu.

16. The answer is D. *(Alberts, 3/e, pp 834–846. Goodman, pp 82–83.)* The fundamental structure of the actin molecule is the same no matter what the function or arrangement in a cell. The stability, arrangement, and functions of actin filaments are dependent on the actin-binding proteins. These proteins have a variety of functions: (1) tropomyosin strengthens actin filaments, (2) fimbrin and villin are actin-bundling proteins, (3) filamin and gelsolin regulate transformation from the sol to the gel state with filamin cross-linking filaments into the gel while gelsolin fragments filaments, (4) members of the myosin II family are responsible for sliding filaments, (5) myosin I (minimyosin) is responsible for movement of vesicles on filaments, and (6) spectrin cross-links the sides of actin filaments to the plasma membrane.

17. The answer is D. *(Alberts, 3/e, pp 601–603. Goodman, pp 121–122. Junqueira, 8/e, pp 32–35. Ross, 3/e, pp 31–32. Stevens, pp 16–17.)* The Golgi apparatus presents two faces: a *cis* face, which is the point of entry of transport vesicles in transit from the RER to the Golgi, and a *trans* face, which is the exit point associated with the maturation of proteins. Both proteins and lipids are transported from the transitional elements of the ER to the Golgi. Packaging is not the sole function of the Golgi. This organelle is also involved in the processing of proteins (e.g., oligosaccharide chains) that was initiated in the RER. There is a specific organization to the Golgi stacks related to the function of specific enzymes. (Golgi processing will be reviewed in later questions that deal with secretion.) Histochemical stains such as acid phosphatase and nucleoside diphosphatase demonstrate that the Golgi apparatus is biochemically compartmentalized.

18. The answer is C. *(Alberts, 3/e, pp 804–805. Goodman, pp 88–94. Junqueira, 8/e, pp 40–42. Stevens, p 20. Widnell, pp 84–86.)* Microtubules are labile cytoplasmic structures capable of very rapid assembly and disassembly. There are a number of drugs that inhibit mitosis. Colchicine inhibits the addition of tubulin molecules to microtubules and therefore induces depolymerization of microtubules. Vinblastine also causes microtubule depolymerization through the formation of paracrystalline aggregates of tubulin. Taxol stabilizes microtubules by binding tightly to them and induces tubulin to form microtubules. Microtubules differ from microfilaments, which are often found in bundles with extensive cross-linking. In comparison, microtubules are often found singly in the cytoplasm.

19. The answer is B. *(Alberts, 3/e, pp 801–803. Goodman, pp 86–88.)* The different types of intermediate filaments all have a similar structural pattern: nonhelical head and tail segments with a helical arrangement in the center of the intermediate filament structure. While there are differences in the way that intermediate filaments interact with microtubules and microfilaments within the cytoplasm, their ropelike arrangement is well suited to their function of providing mechanical stability to the cell and resisting stretch, allowing the cell to respond to tension. Movement is generated by motor proteins such as myosin, dynein, and kinesin. Nucleation of microtubules is carried out by centrosomes. Microtubule-associated proteins (MAPs) stabilize microtubules against disassembly. Microtubules function in organellar transport (e.g., axonal transport).

20. The answer is C. *(Alberts, 3/e, pp 577–582. Goodman, pp 115–116, 118–119. Junqueira, 8/e, pp 31–32.)* The smooth endoplasmic reticulum (SER) is that portion of the endoplasmic reticulum (ER) that lacks ribosomes. It may be separated from the RER by differential centrifugation using sucrose gradients, a technique in which smooth and rough microsomes are formed. Smooth microsomes possess a lower density than rough microsomes and float at a lower sucrose concentration. SER is involved in the regulation of Ca^{2+} during muscle contraction and therefore plays a critical role in this process. SER concentration and arrangement vary from cell to cell. SER is also involved in lipid synthesis. In hepatocytes there is extensive SER, which is involved in detoxification using enzymes such as cytochrome P450. The SER is derived from the RER and most of the membrane components can diffuse freely from RER to SER since they are part of a continuous fluid-mosaic membrane. The enzymes required for SER function are synthesized in ribosomes on the RER, and some enzymes, such as those involved in the degradation of glycogen (e.g., glucose-6-phosphate), are found in both SER and RER. The ribophorins, which are specific integral membrane proteins found in the RER, are a notable exception to these similarities between SER and RER membranes.

21. The answer is B. *(Alberts, 3/e, pp 653–683. Junqueira, 8/e, pp 27–29.)* Mitochondria produce energy that the cell uses in transport and other energy-dependent processes. Cellular energy is stored as ATP, which is synthesized by the phosphorylation of ADP by ATP synthetase. Mitochondria utilize an electron-transport (respiratory) chain that transfers energy from NADH to O_2. This energy is used by mitochondria to establish an electrochemical proton gradient through the pumping of protons from the mitochondrial matrix. The proton pumping action is based upon both pH and membrane potential gradients between the matrix and the inner mitochondrial membrane and represents stored energy that may be converted into ATP. ATP synthetase utilizes the backflow of protons toward the matrix to form ATP and therefore couples oxidative transport through the electron-transport (respiratory) chain with energy storage (ATP).

22. The answer is D. *(Alberts, 3/e, pp 821–825. Goodman, pp 62, 67, 95.)* Actin monomers assemble in vitro in the presence of energy in the form of ATP. Nucleation is the delay that occurs in vitro before actin monomers are polymerized in a specific structural conformation. Assembly continues until the concentration of monomers in solution reaches a critical concentration or state of equilibrium in which association and dissociation balance one another. Below this level, disassembly of the polymer occurs to raise the monomer solution back to the critical level. Treadmilling occurs in vitro for both microfilaments (actin) and microtubules. During actin polymerization, for example, actin monomers are continuously dissociated at the minus end of the polymer and added to the plus end. This results in maintenance of the monomer concentration in solution with a constant length of the filament maintained by transfer of actin molecules from one end to the other of the actin filament.

23. The answer is A. *(Alberts, 3/e, pp 577–579. Goodman, pp 115–116. Junqueira, 8/e, pp 31–32. Ross, 3/e, pp 26–30.)* The endoplasmic reticulum (ER) is found in all eukaryotic cells and may represent more than 50 percent of the total membrane of the cell. It is a membrane-bound organelle that varies in size depending upon the functional activity of the cell. ER may be classified as rough (RER) or smooth (SER). RER is associated with ribosomes involved in the synthesis of proteins for export but also integral membrane proteins and proteins that constitute the membranes of mitochondria and peroxisomes. SER is involved in lipid synthesis and is extensive in cells actively involved in lipid production, such as the cells of the adrenal cortex. It may be induced in other cells such as hepatocytes by the systemic administration of drugs such as phenobarbital. Therefore, in the case of both the SER and RER, the amount of ER varies with the metabolic activity of the cells.

24. The answer is E. (*Alberts, 3/e, pp 576–577.*) Zellweger syndrome is a disease in which peroxisomes are empty because of the failure of the signal system that sorts protein to this organelle. Since the peroxisome lacks a genome or synthetic machinery, it must import all proteins. In the case of Zellweger syndrome, it appears that the defect is in the peroxisomal membrane, but errors or absence of the peroxisomal signal sequence would result in the same symptom. Peroxisomes were first identified in liver and kidney cells, which have large numbers of peroxisomes because of their function in detoxification and waste removal. Peroxisomes protect the cell by removal of H_2O_2 and the detoxification of alcohol through the action of alcohol dehydrogenases. The absence of catalase and other proteins of the peroxisome would result in an inability to detoxify alcohol. The lysosomal degradation system should be independent of peroxisomal function.

25. The answer is E. (*Alberts, 3/e, pp 574–575. Goodman, pp 140–141. Junqueira, 8/e, pp 37–38, 46.*) Peroxisomes, or microbodies, are membrane-bounded organelles that contain oxidases that catalyze substrates of molecular O_2. The resulting peroxide (H_2O_2) is degraded by catalase, which is also localized in peroxisomes. Although peroxisomes are compared with mitochondria, they have a single membrane while mitochondria have a double membrane. In addition, peroxisomes lack the genetic material and ribosomes found in mitochondria. The absence of a genome requires that peroxisomes import all of their proteins and membrane lipids. This process is similar to that which occurs in the importation of proteins to the mitochondria. The peroxisome is similar to a mitochondrion in that it is involved in the generation of energy, although electron-transport chains are absent. β-Oxidation of long-chain fatty acids results in the formation of the substrates for lipid-processing enzymes found in peroxisomes. Peroxisomes are similar to mitochondria in that new organelles form through the growth and fission of existing organelles.

26. The answer is D. *(Alberts, 3/e, pp 823, 826, 917, 928–932. Goodman, pp 83, 268.)* Cytochalasins are potent inhibitors of cell motility and other cellular events that are dependent on actin assembly: cytokinesis, which is carried out by the actin-containing contractile ring; phagocytosis; and formation of lamellipodia. Cytochalasins bind to the plus end of actin filaments and prevent further polymerization. The movement of chromosomes in anaphase of the cell cycle is dependent upon disassembly of microtubules at the kinetochore (anaphase A) and addition at the plus end of the polar microtubules (anaphase B).

27. The answer is E. *(Alberts, 3/e, pp 622, 628–629.)* A vacuolar H^+-proton pump is present in the membranes of most endocytic and exocytic vesicles, including those of the phagosomes, lysosomes, secretory vesicles, and some compartments of the Golgi. Acidification causes concentration of the contents of secretory vesicles, facilitates breakdown of the contents of phagosomes and lysosomes, and is involved in the cleavage of prohormones to their active forms (e.g., proinsulin to insulin).

28–30. The answers are 28-A, 29-B, 30-A. *(Alberts, 3/e, pp 655–658. Goodman, pp 131–132. Junqueira, 8/e, pp 27–29. Ross, 3/e, pp 32–35. Stevens, p 13.)* The large ovoid structure in the figure is a mitochondrion. The structure labeled A is the mitochondrial matrix, or intercristal space. The matrix contains the circular DNA of the mitochondrial genome. Most mitochondrial proteins are encoded for by nuclear DNA, but a small proportion are encoded within the mitochondrial DNA and are synthesized on mitochondrial ribosomes. The matrix also contains the enzymes responsible for the Krebs (citric acid) cycle. Matrix granules, which probably consist of accumulations of calcium ions, have been identified (D).

The structure labeled B is the inner membrane of the mitochondrion, which is highly impermeable to small ions because of the presence of cardiolipin. The inner membrane contains the proteins required for the oxidative reactions of the respiratory transport chain and a transmembrane complex (ATP synthetase) that is responsible for ATP synthesis. This membrane is folded into convolutions called *cristae*. The number of cristae is directly related to the metabolic activity of the cell. The elementary particles that have been identified on the cristae are probably composed primarily of ATP synthetase complexes.

The outer mitochondrial membrane, labeled C, is highly permeable to molecules 10,000 daltons or less because of the presence of porin, a channel-forming protein. This membrane contains enzymes involved in lipid synthesis and lipid metabolism. The outer membrane mediates the movement of fatty acids into the mitochondria for use in the formation of acetyl CoA.

The mitochondrial granules are labeled D and are believed to be divalent cationic binding sites. The intracristal space is labeled E.

31–33. The answers are 31-C, 32-D, 33-C. *(Alberts, 3/e, pp 796–802, 813–815, 831–832. Goodman, pp 83, 86–88, 91–92. Junqueira, 8/e, pp 38–45. Stevens, pp 19–22.)* Microfilaments (thin filaments) are composed of actin, the most abundant protein in cells of eukaryotes. They are involved in cell motility and changes in cell shape. Myosin is the main constituent of the thick filament that binds to actin and functions as an ATPase activated by actin. Intermediate filaments, which are intermediate in diameter (8 to 10 nm) between thin and thick filaments, are of four different types. Type I is composed of the acidic, neutral, and basic keratins (also known as the *cytokeratins*) and is found specifically in epithelial cells. Type II intermediate filaments are composed of vimentin, desmin, or glial fibrillary acidic protein. Vimentin is found in cells of mesenchymal origin, desmin in muscle cells, and glial fibrillary acidic protein primarily in astrocytes. Type III intermediate filaments are neurofilament proteins found in neurons. Type IV intermediate filaments consist of nuclear lamins A, B, and C and are associated with nuclear lamina of all cells. Microtubules are structures composed of tubulin, the principal protein in the composition of the axoneme (the core of the cilium or flagellum). Kinesin is an ATPase that hydrolyzes ATP to ADP using the resulting energy to move vesicles unidirectionally along the microtubules (e.g., from the perikaryon of a neuron to the axon terminus). In undifferentiated, embryonic cells there are few intermediate filaments, which increase with the degree of differentiation of the cells. In damaged cells, there is a collapse in the structure of the intermediate filaments, but the structure reforms when cell repair is complete. Spectrin is not involved in the aforementioned cellular processes.

34–35. The answers are 34-A, 35-D. *(Alberts, 3/e, pp 491–495, 620–621, 796–798. Goodman, pp 83–86, 128–129. Widnell, p 82.)* Clathrin is an important protein that forms the coating of coated pits and vesicles involved in endocytosis and the retrieval of membrane following exocytosis. Intermediate filaments are important cytoskeletal elements with some specificity that depends upon the origin of the cells in question. Vimentin appears to be specific for cells of mesenchymal origin, such as fibroblasts and chondrocytes. Actin is the protein found in thin filaments. It is also a cytoskeletal component found in the cytoplasm of red blood cells and other eukaryotic cells. Spectrin heterodimers form tetramers that interact with actin and provide flexibility and support for the membrane. The protein ankyrin "anchors" the band 3 protein to the spectrin-membrane skeleton. This connection is often described as the indirect binding of band 3 protein to the cytoskeleton (spectrin tetramers) of the red blood cell. The band 3 protein is known to be the anion transport protein of the red blood cell.

36–38. The answers are 36-E, 37-C, 38-D. *(Alberts, 3/e, pp 601–602. Goodman, pp 121–122, 125.)* The Golgi plays an important role in the processing of proteins for secretion. It is divided into four regions: *cis*-face, medial compartment, *trans*-face, and the *trans*-Golgi network (TGN). Transitional elements are derived from the smooth endoplasmic reticulum (SER) and carry proteins and lipids from the endoplasmic reticulum to the *cis*-face of the Golgi. The *cis*-face of the Golgi receives the transitional elements and participates in phosphorylation (e.g., in the synthesis of lysosomal oligosaccharides). The medial compartment is responsible for the removal of mannose and the addition of *N*-acetylglucosamine. The *trans*-face is responsible for the addition of sialic acid and galactose. The TGN serves as a sorting station for proteins destined for various organelles, including the plasma membrane, and protein for export from the cell. Golgi-derived transport and secretory vesicles bud off from the TGN.

Cell Biology: Intracellular Trafficking

DIRECTIONS: Each question below contains five suggested responses. Select the **one best** response to each question.

39. During endocytosis, low-density lipoprotein (LDL) is dissociated from its receptor in the

(A) clathrin-coated vesicle
(B) endosome
(C) non-clathrin-coated vesicle
(D) lysosome
(E) transport vesicles

40. Binding of epidermal growth factor (EGF) to its receptor results in the formation of a ligand-receptor complex (i.e., EGF–EGF receptor) and subsequently a decrease in cell surface receptors. This occurs through

(A) decreased transcription of EGF receptor genes
(B) decreased translation of EGF receptor message
(C) loss of EGF–EGF receptor complexes in lysosomes
(D) increased degradation of EGF receptor in coated pits
(E) increased recycling of receptor to endosomes

41. Transferrin is a serum glycoprotein responsible for the transport of iron from the blood to organs such as the liver and small intestine. Which of the following occurs in the receptor-mediated pathway for transferrin?

(A) Phagocytosis of iron
(B) Binding of ferrous ions to coated pits with subsequent formation of coated vesicles
(C) Dissociation of ferrous ions from the transferrin-receptor complex in the lysosome
(D) Membrane shuttling of transferrin bound to its receptor, allowing escape from lysosomes
(E) Dissociation of transferrin from the transferrin receptors in lysosomes

42. Phagocytosis may be characterized by which of the following statements?

(A) It involves the uptake of cellular debris in large endocytic vesicles
(B) It is a generalized membrane response and is not localized to specific regions
(C) It is a constitutive process
(D) It is important in digestive processes in mammals
(E) It involves fluid uptake by small vesicles

43. The synthesis and transport of phospholipid differ from those of protein in which of the following ways?

(A) Steps in synthesis are catalyzed by enzymes in the endoplasmic reticulum
(B) Transport vesicles leave the endoplasmic reticulum and supply mitochondria and peroxisomes with newly synthesized phospholipids
(C) Membrane vesicles transport newly synthesized phospholipid to the Golgi and lysosomes
(D) Translocation occurs from the cytosolic to the luminal face of the endoplasmic reticulum
(E) Carbohydrates are added during transport through the intracellular membrane system

44. The diversity in oligosaccharides is produced

(A) in the *trans*-Golgi
(B) in the *cis*-Golgi
(C) by *en bloc* transfer
(D) by selective removal of glucose and mannose from the core oligosaccharide
(E) at the time of insertion in the membrane or secretion

45. Which of the following rough endoplasmic reticulum events occurs during synthesis of protein for export?

(A) A sequence on the C terminus of a newly synthesized peptide recognizes the ER membrane
(B) An N-terminal sequence is cleaved as a peptide is translocated across the ER membrane
(C) A docking protein binds to the signal peptide as it emerges from the ribosome
(D) A signal recognition particle is sequestered in the ER membrane
(E) Binding of the signal peptide causes an immediate stimulation of translation

Questions 46–47

The figure below represents the signal transduction pathways initiated after ligand binding to the β-adrenergic receptor on the cell membrane. This binding leads to stimulation of secretory granule release from a cell such as a pancreatic or salivary acinar cell.

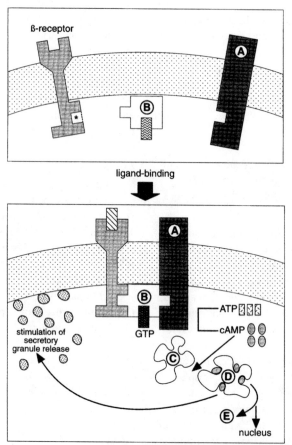

From Avery JK: *Oral Development and Histology,* 2/e. Thieme Medical, 1994, with permission.

46. The molecule labeled A is

(A) adenylate cyclase
(B) guanylate cyclase
(C) muscarinic receptor
(D) activated cAMP kinase
(E) the GTP-binding protein (G_S)

47. Which of the following statements regarding the molecule labeled B is true?

(A) It is the inactive cAMP kinase
(B) It lacks GTPase activity
(C) It inactivates adenylate cyclase
(D) It is bound to GTP in the inactive state
(E) It is the stimulatory G protein (G_S)

48. Chaperonins function to

(A) ensure correct folding of cyto-solic proteins
(B) control the docking of the signal peptide with its receptor on the rough endoplasmic reticulum
(C) stabilize microtubules
(D) serve as a start-transfer signal in the membrane of the endoplas-mic reticulum
(E) mediate the selective transport of membrane receptors

49. Lysosomal hydrolases are tar-geted to the lysosomes by

(A) O-linked oligosaccharides
(B) mannose-6-phosphate added to N-linked oligosaccharides
(C) clathrin-coating of the secretory vesicles
(D) non-clathrin-coating of the secretory vesicles
(E) sialic acid residues

Questions 50–51

The figure below is a diagram of the phosphoinositide (PI) cycle and related regulatory processes.

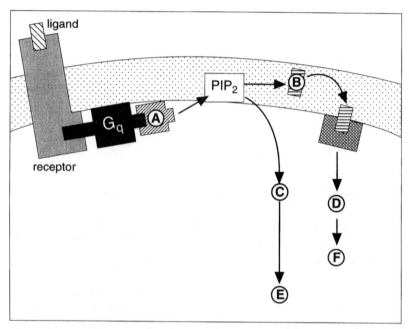

From Avery JK: *Oral Development and Histology,* 2/e. Thieme Medical, 1994, with permission.

50. The interaction of a ligand with the receptor leads to activation of the membrane-associated molecule labeled A, which is

(A) PI-specific phospholipase C
(B) diacylglycerol (DAG)
(C) inositol triphosphate (IP$_3$)
(D) protein kinase C
(E) inhibitory G protein (G$_i$)

51. The function of molecule A is

(A) stimulation of G$_i$ activity
(B) stimulation of G$_s$ activity
(C) hydrolysis of PIP$_2$ to form DAG and IP$_3$
(D) formation of kinase C and PI-phosphate
(E) formation of GTP and ATP

52. Which of the following statements correctly characterizes secretory vesicles?

(A) They bud off from the rough endoplasmic reticulum

(B) They are primarily involved in the constitutive secretory pathway

(C) They contain primarily transmembrane proteins for addition to the plasma membrane

(D) They may be coated with clathrin and its associated coat proteins

(E) They serve only as transport structures to the cell membrane; the secretory contents are not modified in these structures

53. The mechanism involved in nuclear importation of protein differs from other organelle-import mechanisms in

(A) the bidirectional nature of the transport

(B) the biochemical structure of the phospholipid bilayers in the nuclear membrane versus organellar membranes

(C) the absence of a specific signal sequence

(D) passage through an aqueous pore rather than through the membrane

(E) the lack of a requirement for energy

DIRECTIONS: Each numbered question or incomplete statement below is NEGATIVELY phrased. Select the **one best** lettered response.

54. There is a family of diseases related to absence or defects of LDL receptors. Using your knowledge of endocytosis, which of the following would NOT be related to defective function of LDL receptors?

(A) Reduced blood cholesterol levels
(B) Abnormality of the extracellular domain of the LDL receptor, resulting in failure to bind LDL
(C) Abnormality of the cytoplasmic domain of the LDL receptor
(D) Failure of LDL to enter coated pits
(E) Normal numbers of receptors, but failure of LDL to internalize

55. Glycosylation events that occur in the Golgi include all the following EXCEPT

(A) transfer of oligosaccharides by the action of glycosyl (oligosaccharyl) transferase activity
(B) phosphorylation of oligosaccharides on lysosomal proteins
(C) removal of mannose
(D) addition of galactose and sialic acid
(E) polymerization of glycosaminoglycan chains and subsequent sulfation

56. *O*-linked glycosylation *differs* from *N*-linked glycosylation in all the following ways EXCEPT

(A) the location of the process
(B) the way in which oligosaccharides are added
(C) the position of the oligosaccharide addition
(D) the involvement of oligosaccharide protein transferases
(E) the presence of the required enzymes on the luminal side of the cisternal structure

57. Constitutive and regulated secretion *differ* in all the following ways EXCEPT

(A) the involvement of secretory vesicles
(B) pathways of transport from the *trans*-Golgi
(C) the effects of secretagogues on secretory rate
(D) the content of the vesicles
(E) domain selectivity

58. All the following are true of the pathway labeled E in the figure below, which illustrates cell secretion, EXCEPT

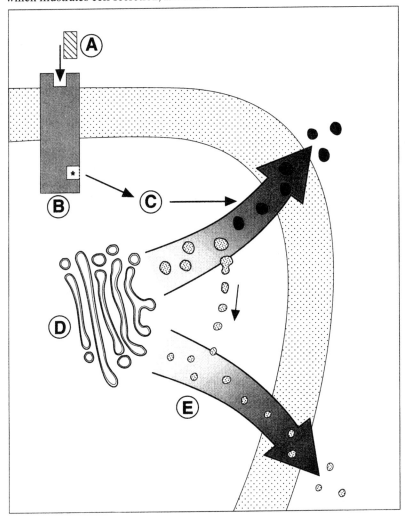

From Avery JK: *Oral Development and Histology,* 2/e. Thieme Medical, 1994, with permission.

(A) it represents the stimulated release of vesicles
(B) it represents the transport of protein constituents of the apical membranes to their destinations
(C) it represents products synthesized by the endoplasmic reticulum with transport from the Golgi
(D) it is utilized in the secretion of extracellular matrix proteins
(E) it represents a nonselective default pathway

59. All the following are true of the signal hypothesis EXCEPT

(A) it involves the translation of a signal sequence of transfer RNA
(B) it explains the translocation of newly synthesized peptides into the cisternae of the rough endoplasmic reticulum
(C) it explains the targeting of transmembrane, lysosomal, and exportable protein
(D) it explains why the addition of an amino-terminal leader peptide to a protein that normally lacks a leader results in translocation across the ER membrane
(E) it explains why microsomal membranes protect newly synthesized proteins from protease activity

60. All the following statements are true of plasma membrane proteins EXCEPT

(A) they are initially inserted into the phospholipid bilayer in the secretory vesicle
(B) they may span the phospholipid bilayer multiple times
(C) the transmembrane portion of the protein functions as a "stop" signal before complete translocation of the protein
(D) many cells have polarized sorting of these proteins in the *trans*-Golgi network
(E) from the Golgi to the plasma membrane these proteins follow a constitutive secretory pathway

61. Inclusion cell disease is a recessive genetic disease in which all the hydrolytic enzymes are missing from the patient's cells but are found in the blood. The etiology is related to a cellular sorting malfunction. All the following would be expected in inclusion cell disease EXCEPT

(A) normal expression of the structural genes encoding the hydrolases
(B) missorting of lysosomal enzymes to the secretory pathway
(C) excessive phosphorylation of lysosomal enzymes
(D) absence of a phosphotransferase that adds mannose-6-phosphate groups to lysosomal hydrolases
(E) accumulation of undigested substrates in the cytoplasm

62. All the following steps in protein processing occur in association with the rough endoplasmic reticulum EXCEPT

(A) formation of disulfide bonds
(B) segregation of protein for export
(C) final processing of oligosaccharides
(D) removal of the signal peptidase
(E) core glycosylation of asparagine residues in glycoproteins

Cell Biology: Intracellular Trafficking

Answers

39. The answer is B. (*Alberts, 3/e, pp 621–624. Goodman, pp 127–130. Ross, 3/e, p 37. Widnell, pp 123–124.*) Low-density lipoprotein (LDL) is the form in which most cholesterol is transported in the blood; cellular uptake of LDL is the classic example of receptor-mediated endocytosis. Receptor-mediated endocytosis is a selective process initiated by the binding of a macromolecule (ligand) to a receptor on the plasma membrane. The receptors are bound to clathrin-coated pits, but the ligand is only directly bound to its cell surface receptor. LDL in the blood is bound to LDL receptors, which are found in the membrane. The LDL receptors (ligand-receptor complexes) are incorporated into the cell by the formation of coated pits. The coated pits bud off from the surface to form coated vesicles. The acidic environment of the endosome results in the cleavage of the ligand from its receptor (i.e., LDL from its receptor). The LDL receptors are recycled to the membrane for additional exposure to LDL, and the LDL in the endosome is transferred to lysosomes, where it is broken down to cholesterol. The recycling of receptors through the fusion of endosomes with vesicular structures occurs in a compartment sometimes called *CURL (compartment for uncoupling of receptor and ligand).*

40. The answer is C. (*Alberts, 3/e, pp 623–624. Goodman, p 131. Stevens, pp 9–10. Widnell, pp 117–120.*) In the example of epidermal growth factor (EGF)—a peptide that is synthesized by the salivary glands, Brunner's glands (submucosal glands of the duodenum), and other organs and serves as a mitogen for various tissues and organs—the steps in receptor-mediated endocytosis are similar to those for other ligands such as LDL. These steps include (1) binding of EGF to EGF receptor, (2) accumulation of EGF–EGF receptor complexes in coated pits, (3) entry of the complexes into the cell through coated pits, and (4) transformation of the pits into vesicles, which subsequently lose their coats before endosomal fusion. EGF binding does differ from endocytosis of other molecules such as transferrin and LDL in that binding of EGF to its receptor is required before EGF receptors are capable of accumulation in coated pits. More importantly, in the case of the EGF receptor, EGF binds to receptor but the complex is not dissociated at the 5.0 to 5.5 pH of the endosome; the result is transfer of the complex to the lysosomes with the eventual degradation of EGF receptors in lysosomes and a reduction in the number of EGF receptors on the surface. The decrease in the number of receptors is known as down-regulation of receptors and does not require transcriptional or translational responses.

41. The answer is D. (*Alberts, 3/e, p 624. Goodman, pp 130–131.*) Transferrin offers an example of endocytosis that differs somewhat from the LDL system. In the case of transferrin, iron-free transferrin (apotransferrin) binds ferrous ions in the blood. The iron-bound molecule (transferrin) attaches to transferrin receptors on the cell surface, initiating the receptor-mediated pathway through coated pits and vesicles. In the endosome, there is dissociation of the iron molecules from the transferrin–transferrin receptor complex. Iron is released into the cytosol for cellular utilization, and the ligand-receptor complex is recycled to the membrane and subsequently released from the cell. In the neutral pH of the extracellular milieu, the ligand dissociates from its receptor and is ready to bind additional ferrous ions.

42. The answer is A. (*Alberts, 3/e, pp 618–619. Goodman, pp 127–128. Widnell, pp 119–120.*) Phagocytosis, endocytosis (receptor-mediated and fluid-phase), and autophagy are the three pathways of materials to the lysosomes. Phagocytosis is a form of ingestion and therefore is one of the two types of endocytosis; the other type is pinocytosis (fluid uptake by small vesicles). Phagocytosis represents the uptake of microorganisms and debris by specific cells of the body. Macrophages are phagocytic cells found throughout the body, e.g., microglia in the brain, Kupffer cells in the liver, and Langerhans cells in the skin. Other cells such as neutrophils, eosinophils, and basophils are involved in phagocytosis. Phagocytosis is not involved in digestion in the small intestine, since a variety of enzymes present in the lumen as well as brush border enzymes degrade nutrients to smaller molecules so that uptake is facilitated in the small bowel. Phagocytosis is the process of ingestion of large structures by the cell. The process is triggered by the presence of certain particles at the cell surface; it is not a constitutive mechanism and only occurs in regions of the cell in contact with the substance to be ingested. Agents such as cytochalasin, which disrupt actin polymerization, block phagocytosis by prevention of the formation of macrophage pseudopods and indicate the importance of the cytoskeleton in phagocytosis. Antibodies are the best example of a stimulus for the phagocytic process. Antibodies bind to the surface of antigens, targeting them for ingestion by macrophages and other phagocytic cells. Adhesion in this case is the recognition and binding of IgG to Fc receptors on the macrophage. The binding of ligand to receptor is probably coupled to the initiation of changes in the cytoskeleton that lead to phagocytosis. Membrane-zippering is involved in phagocytosis since continuous contact of the antibodies with the Fc receptors is required for the pseudopod to surround the particle to be ingested. If the IgG molecules are limited to a specific region of the antigen, then there is contact, but no ingestion. Inert as well as biologic materials are ingested by macrophages. Therefore, latex beads, india ink, and trypan blue are phagocytosed by macrophages when injected into the body or provided in the medium surrounding cells in vitro.

43. The answer is D. *(Alberts, 3/e, pp 592–594, 604–605. Goodman, pp 118–120. Widnell, p 115.)* Phospholipid synthesis occurs in the endoplasmic reticulum (ER). Phosphatidylcholine (lecithin) is the major phospholipid synthesized by the cell. Initial formation of phosphatidic acid occurs in the cytosolic half of the ER membrane. Additionally, enzymes required for the processing and synthesis of phosphatidic acid to form diacylglycerol and phosphatidylcholine are also located on the cytosolic half of the membrane. Newly formed molecules of phospholipid are added to the cytosolic half of the ER membrane; however, enzymes called *flippases* are responsible for the flip-flop, or translocation, of phospholipid to the luminal side of the membrane. These enzymes are specific for different head groups of the phospholipids. This process is one point of difference between phospholipid and protein synthesis. The mitochondria and peroxisomes are not part of the continuous internal membrane–plasma membrane communication system. Therefore, transport vesicles do not transport phospholipid to these organelles. There is a system of water-soluble carrier proteins, known as *phospholipid exchange proteins,* which are responsible for the transport of phospholipid to the mitochondria and peroxisomes. This process is believed to occur through a random exchange mechanism based on concentration gradients. Nonvesicular exchange occurs most readily with lipids in the cytoplasmic half of the bilayer where the phospholipid can be easily bound to carrier proteins, which can then travel to another cytoplasmic site and transfer the lipid to another membrane (e.g., mitochondrial membranes). Transport within the intracellular membrane system occurs by vesicular transport for both protein and lipid. Carbohydrates are added to lipids in the Golgi. Carbohydrate modifications of proteins occur in both the rough ER and the Golgi. Therefore, in both glycolipids and glycoproteins, carbohydrate modifications occur in their maturational journey through the intracellular membranes.

44. The answer is D. *(Alberts, 3/e, pp 589–591, 604–606. Goodman, pp 117–119. Widnell, pp 107–109.)* Glycosylation of proteins occurs in the rough endoplasmic reticulum (ER) by the *en bloc* transfer of a common oligosaccharide consisting of *N*-acetylglucosamine, mannose, and glucose.

The diversity in oligosaccharides is produced by selective removal of glucose and mannose from the core oligosaccharide. This trimming process occurs in the rough ER before reaching the Golgi. The addition of oligosaccharide is *N*-linked or asparagine-linked. The oligosaccharide is transferred from dolichol, which binds oligosaccharide within the ER membrane. The enzyme for the transfer of the oligosaccharide is located on the luminal surface of the ER, which prevents exposure to cytosolic proteins.

45. The answer is B. (*Alberts, 3/e, pp 582–586. Goodman, p 116. Widnell, pp 107–109.*) In the function of the rough endoplasmic reticulum (RER), a presequence on the 3′end of the AUG initiation codon is translated as an N-(amino)-terminal presequence that recognizes the ER membrane and eventually leads to the translocation of the peptide across the ER membrane. This recognition is accomplished through the signal recognition particle (SRP), which cycles between the ER membrane and the cytosol. The SRP binds to the signal peptide as the peptide emerges from the ribosome and induces an immediate delay in translation until the ribosome interacts with a docking protein, also known as an *SRP receptor,* in the ER membrane. After the SRP-bound ribosome attaches to the ER membrane via the docking protein, translation continues with displacement of the SRP for subsequent recycling and translocation of the peptide across the ER membrane.

46–47. The answers are 46-A, 47-E. (*Alberts, 3/e, pp 734–738. Avery, 2/e, p 374. Goodman, pp 244–246. Junqueira, 8/e, pp 25–27.*) The figure illustrates the response of the β-adrenergic receptor to ligand binding. β-Receptors function to mediate the tissue effects of epinephrine and norepinephrine in the body. They also respond to pharmacologic agents such as isoproterenol, a β-adrenergic agonist drug. The subsequent signal transduction following ligand binding involves a specific member of the G-protein family of cell surface receptors and a chain of intracellular mediators, also known as *second messengers. G proteins* is shorthand for *guanosine-triphosphate (GTP)–binding regulatory proteins.* Those G proteins associated with increasing cAMP levels in the cell are known as stimulatory G proteins (G_S) because of their role in enzyme activation. In the inactive state G_S is bound to GDP. When isoproterenol or another ligand binds to the β-receptor, a G_S binding site is exposed and the G_S protein (B in the figure) binds to the β-receptor. The resulting complex is capable of binding GTP in exchange for GDP, activating the G protein. A subunit of the activated G_S protein activates adenylate cyclase (A in the figure).

Three different polypeptide chains compose G proteins. For this reason they are often called *trimeric G proteins.* The three subunits are α, β, and γ. The α-subunit of the G_S protein ($α_S$ subunit) exchanges GDP for GTP in response to stimulation in the form of binding to a ligand-activated receptor. This subunit is also responsible for binding to adenyl cyclase. When this occurs, GTPase activity is increased, which results in a short activation time (less than 1 min) for the complex and allows recycling of the subunits to the inactive state. The inactive cAMP kinase is labeled C in the figure. It is the phosphorylating action of cAMP-dependent protein kinase (kinase A) that affects many aspects of intracellular metabolism and function. In addition to the effect on phosphorylation (E) that stimulates exocytosis and other cellular events, the protein phosphorylation also induces nuclear changes including transcriptional events. The star in the figure delineates the site of ligand-binding–induced conformational change, exposing the G_S-binding site.

48. The answer is A. *(Alberts, 3/e, pp 571–572. Wynn, J Lab Clin Med 124: 31–36, 1994.)* The chaperonins are proteins that regulate the unfolding of cytosolic proteins. They are members of the heat shock protein family (e.g., hsp 70). The chaperonins assist with the translocation of proteins across internal membranes of the cell (e.g., mitochondria) by binding precursor proteins in their unfolded state during the process of movement across the membrane. They do not function in the docking of the signal peptide or as the start-transfer signal in translocation of the internal membrane of the endoplasmic reticulum. Clathrin-coated vesicles are responsible for the selective transport of membrane receptors.

49. The answer is B. *(Alberts, 3/e, pp 613–616. Goodman, p 123. Widnell, p 112.)* The targeting of lysosomal enzymes is based on the addition of mannose-6-phosphate to *N*-linked oligosaccharides. The *N*-linked oligosaccharides are recognized by mannose-6-phosphate receptors, transmembrane proteins that become concentrated in the coated vesicles that bud from the *trans*-Golgi network and are observed in the endolysosomes.

50–51. The answers are 50-A, 51-C. *(Alberts, 3/e, pp 744–745. Avery, 2/e, p 375. Goodman, pp 248–249.)* Phosphoinositides are important intracellular second messengers. The so-called phosphatidylinositol (PI) cycle is based on the formation of PI-bisphosphate (PIP_2) in the inner leaflet of the plasma membrane. The breakdown of PIP_2 leads to the formation of the key functional agents of the PI cycle. The process begins with the binding of a ligand to its G-protein–linked receptor on the cell surface. In this case, the trimeric G protein was once called G_p, but is now known as G_q. It functions to activate a phosphoinositide-specific phospholipase C.

PI-specific phospholipase C hydrolyzes PIP_2 to form DAG and IP_3. These two molecules function differently to regulate intracellular function. IP_3 functions in the mobilization of calcium while DAG activates protein kinase C, leading to multiple phosphorylations of cytosolic proteins.

DAG (B in the figure) is responsible for activation of protein kinase C (so-called because of its Ca^{2+} dependency), which is labeled D. The protein kinase C phosphorylates specific serine and threonine residues in the cytosol and it functions in many cells to alter gene transcription. In contrast IP_3 functions to mobilize Ca^{2+} (E) by binding to IP_3-gated Ca^{2+}-release channels in the membranes of the endoplasmic reticulum. The two intracellular messenger pathways do interact in that elevated Ca^{2+} translocates protein kinase C from the cytosol to the inner leaflet of the plasma membrane.

52. The answer is D. (*Alberts, 3/e, pp 626–628. Goodman, pp 124–127. Widnell, pp 105–107.*) Secretory vesicles bud off from the *trans*-Golgi network (TGN) of the Golgi apparatus and transport primarily exportable proteins to the plasma membrane for exocytosis. They are responsive to secretory stimuli and secretagogues and therefore function within the regulated secretory pathway. In contrast, the constitutive secretory pathway is involved in the continual transport of transmembrane proteins and lipids to the plasma membrane for replacement of membrane proteins. The secretory vesicles serve to concentrate the secretory product and in the case of some peptides to convert precursors to active forms (e.g., proinsulin to insulin). Both the concentration process and the conversion of precursors to active forms of peptides appear to involve acidification through a proton pump mechanism in the membrane of the secretory vesicles. When secretory vesicles bud off from the TGN, they may be coated with clathrin and its associated coat proteins. The clathrin-coated vesicle is denuded of the clathrin coating during the concentration process within the secretory vesicles.

53. The answer is D. (*Alberts, 3/e, pp 564–568.*) Nuclear protein signaling is required for the entry of appropriate proteins into the nucleus. Nuclear import signals are specific for nuclear proteins and open an aqueous pore for the passage of nuclear import proteins. Energy appears to be required for the import of nuclear proteins as it is for other organelle transport processes. The import process is unidirectional in both cases. The structure of the membranes, except for the presence of pores, is very similar. The presence of an aqueous pore is the primary difference in the nuclear importation process compared with that of other organelles.

54. The answer is A. (*Alberts, 3/e, pp 621–622. Goodman, pp 129–130. Widnell, p 128.*) The LDL receptor has been studied by Brown and Goldstein, who received the Nobel Prize for their important cellular biologic research related to hypercholesterolemia. In this disease the absence of LDL receptors or presence of defective LDL receptors results in elevated blood cholesterol levels (hypercholesterolemia). The abnormality may occur in the extracellular domain of the LDL receptor, resulting in a failure to bind LDL. Alternatively, the cytoplasmic domain of the LDL may be defective, resulting in failure of the receptors to form in the coated regions of the plasma membrane and the subsequent failure of LDL receptors to enter coated regions of the plasma membrane. This may result in the failure of LDLs to enter coated pits and internalize despite normal numbers and extracellular domain structure of the LDL receptors. Failure of entry or internalization of LDL, a complex of cholesterol molecules and a lipid monolayer, results in elevated blood levels of cholesterol. The ultimate effect for the patient is severe atherosclerosis and coronary artery pathology.

55. The answer is A. *(Alberts, 3/e, pp 589–591, 604–606. Goodman, pp 117–119.)* In the formation of *N*-linked oligosaccharides, *en bloc* transfer of a single species of oligosaccharide occurs by the action of transferases in the endoplasmic reticulum. Trimming of the oligosaccharide also occurs in the endoplasmic reticulum. Events that occur in the Golgi include the phosphorylation of oligosaccharides on lysosomal proteins, removal of mannose, addition of galactose and sialic acid, addition of *N*-acetylglucosamine, and polymerization of glycosaminoglycan chains with subsequent sulfation. This last event is involved in the formation of proteoglycans, which are also glycosylated in the Golgi. The sugars that are incorporated into glycosaminoglycans are sulfated in the Golgi as well.

56. The answer is D. *(Alberts, 3/e, pp 606–607. Widnell, p 109.)* *N*-Linked oligosaccharides are the most common oligosaccharides found in glycoproteins. They contain sugar residues linked to the NH_2 amide nitrogen of asparagine. *O*-Linked oligosaccharides are less common than the *N*-linked species. Linkage of the sugar residues is to hydroxyl groups on the side chains of serine and threonine. *O*-Linkage is catalyzed by glycosyltransferase enzymes in the Golgi, not the rough endoplasmic reticulum. However, in both the ER and the Golgi apparatus, the enzymes are located on the luminal side of these cisternal structures. The addition occurs sugar by sugar rather than *en bloc* as occurs in the rough endoplasmic reticulum for the *N*-linked oligosaccharides.

57. The answer is A. *(Alberts, 3/e, pp 626–628. Goodman, pp 114, 124, 126.)* Constitutive secretion and regulated secretion represent the two secretory pathways. Regulated secretion requires a response to a secretagogue, while constitutive secretion is a default pathway that continues in the presence or absence of a secretagogue. The secretory process is similar for products destined for the two secretory processes until they reach the *trans*-Golgi. In the *trans*-Golgi, signals are required for signaling to lysosomes (mannose-6-phosphate receptor) for the regulated pathway, in contrast to the absence of signals required for the default-constitutive pathway. It is also true that both forms of secretion involve vesicular transport, although unregulated membrane fusion occurs in the constitutive pathway. This is in contrast to the regulated membrane fusion that defines the regulated pathway. Secretory vesicles (also known as *secretory granules*) contain secretory products to be released by secretagogue stimulation. Secretory vesicles are clathrin-coated at the time that they bud off from the *trans*-Golgi. They lose their clathrin-coating and condense the secretory product by acidification by a proton-pump system. Secretory granules are stored and subsequently released upon secretagogue stimulation. Constitutive secretion demonstrates no domain specificity; regulated secretory processes are capable of directing products to apical or basolateral domains as appropriate. While the same products may be found in the constitutive and regulated pathways, generally products for the constitutive pathway include newly synthesized plasma membrane lipids and proteins and newly synthesized soluble proteins and extracellular matrix proteins.

58. The answer is A. (*Alberts, 3/e, pp 626–628. Goodman, pp 124–127.*) The figure represents the differences between the two types of secretion. Regulated secretion (C) shows the recognition of a receptor (B) for its ligand (A), resulting in the release of secretion in response to the stimulus of secretagogue binding to receptor. One of the best examples is the binding of cholecystokinin (CCK) to a pancreative acinar cell, resulting in the release of pancreatic enzymes, which constitute the pancreatic juice. The vesicles that bud from the *trans*-Golgi network (D) in this system are clathrin-coated and contain a receptor involved in the concentration of secretory product, which normally occurs before release. The nonselective default pathway (E) represents a method for shuttling proteins such as integral membrane proteins and lipids to the apical and basolateral membranes of the cell. It represents unstimulated release compared with the secretagogue-mediated secretion that occurs in the regulated pathway. (Also, see discussion for question 57.)

59. The answer is A. (*Alberts, 3/e, p 582. Junqueira, 8/e, pp 31–32. Stevens, pp 14–15. Widnell, pp 107–109.*) The signal hypothesis is the basis of the targeting of transmembrane, lysosomal, and exportable proteins across the ER membrane. This process is dependent on an amino-terminal signal leader (prepeptide) sequence that recognizes a receptor on the ER membrane. This is the key event in the segregation of noncytosolic proteins in the ER cisternae away from the cytoplasm. Experimentally, it has been possible to add signal sequences to proteins that normally do not have leader sequences. The result is a protein that is translocated across the ER membrane. Much of the research on the signal hypothesis has been carried out using in vitro cell-free translation systems. Using this experimental design, it was found that two conditions were required to prevent degradation of newly synthesized peptides by proteases: the presence of a signal sequence and the presence of microsomes. The translocation of the newly formed peptides across the ER membrane results in protection from protease activity. The tRNAs are responsible for the conversion of mRNA nucleotide sequence into a correct protein sequence by matching the anticodon of the tRNA to the codon on the mRNA. The message is present in the mRNA, and in any case these functions are not part of the signal hypothesis, which relates to the translocation of protein into the RER cisternae.

60. The answer is A. *(Alberts, 3/e, pp 584–589. Widnell, pp 113–114.)* Many plasma membrane proteins are transmembrane proteins. Some of these are multiple-pass transmembrane proteins, which span the lipid bilayer more than once. These proteins are initially inserted into the phospholipid bilayer in the rough endoplasmic reticulum. As they are being translocated into the cisternal space at the level of the endoplasmic reticulum, a hydrophobic (transmembrane) portion of the molecule serves as a stop transfer signal, which results in arrest of transport. Many cells are polarized with specific proteins targeted to the apical and basolateral domains of the membrane. Targeting is established in the *trans*-Golgi network, although there are exceptions in which some proteins initially targeted to the basolateral membrane are subsequently shuttled to the apical membrane. Plasma membrane proteins follow the same pathway as exportable, constitutively released proteins through vesicular transport to the membrane.

61. The answer is C. *(Alberts, 3/e, pp 617–618. Goodman, pp 123–124.)* In inclusion cell disease, there is an absence or deficiency of *N*-acetylglucosamine phosphotransferase. This results in missorting of lysosomal enzymes to the secretory pathway since the absence of phosphorylation in the *cis*-Golgi prohibits segregation of lysosomal enzymes that normally occurs in the *trans*-Golgi through the action of mannose-6-phosphate receptors. Lysosomal enzymes are secreted into the bloodstream, and undigested substrates build up within the cells. There is normal expression of the genes encoding the hydrolases, but a misdirection of the intracellular sorting signal for these hydrolytic enzymes.

62. The answer is C. *(Alberts, 3/e, pp 582–586. Goodman, p 116. Junqueira, 8/e, pp 31–33. Widnell, pp 107–109.)* Protein processing begins with protein synthesis on attached polysomes (polyribosomes) in the cytoplasmic matrix. The rough endoplasmic reticulum (RER) is responsible for segregation of exportable protein from the cytosol through the signal peptide and the action of the signal recognition particle, the docking protein (signal recognition particle receptor), and the ribophorins. Enzymes in the RER are responsible for formation of disulfide bonds, which assist in the segregation of proteins in the cisternae of the RER. Cotranslational events that occur in the RER include the removal of the signal peptide by a signal peptidase and core glycosylation of asparagine residues in glycoproteins. The final processing of oligosaccharides, however, is a Golgi event.

Cell Biology: Nucleus

DIRECTIONS: Each question below contains five suggested responses. Select the **one best** response to each question.

63. The structure labeled in the electron micrograph of a nucleus below

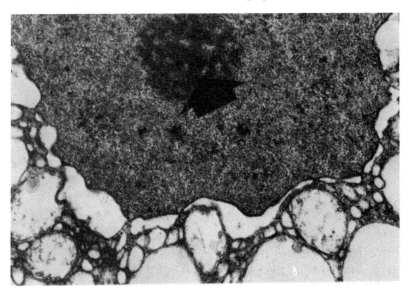

(A) is a homogeneous structure in most cells
(B) contains an organizer composed of DNA
(C) is the site of complete ribosomal assembly and maturation
(D) synthesizes the ribosomal proteins
(E) is of uniform size and number in different cells

64. The dividing cell shown in the electron micrograph below is involved in the stage of the cell cycle characterized by which of the following statements?

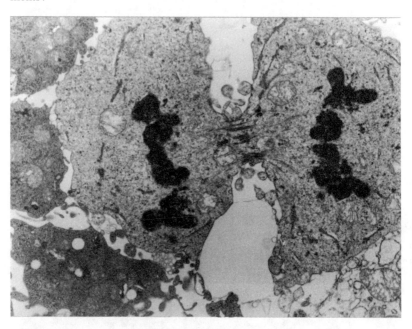

(A) It occurs before nuclear division during a normal cell cycle
(B) It is accomplished by the contraction of a ring composed of cytoskeletal elements
(C) It occurs through the lengthening of kinetochore microtubules
(D) It occurs through the shortening of polar microtubules
(E) It is blocked by antitubulin antibodies

---

65. The function of the nucleosomes is to

(A) package genetic material in a condensed form
(B) transcribe the DNA
(C) form pores for bilateral nuclear-to-cytoplasmic transport
(D) form the nuclear matrix
(E) hold together adjacent chromatids

66. Which of the following proteins binds to membrane proteins and serves as a scaffold to support the nuclear envelope?

(A) Lamins
(B) Actin
(C) Microtubules
(D) Chaperonins
(E) Porins

67. The characteristic mass of heterochromatin lying against the nuclear membrane in nucleated cells is known as the *Barr body.* Which of the following statements is true of the Barr body?

(A) It represents the inactivated Y chromosome of males that remains condensed during interphase

(B) Its absence in a buccal smear chromatin test indicates definitively that the patient is female

(C) It would have an identical appearance in a buccal smear chromatin test in a patient with Turner's syndrome and in a normal female

(D) It is found exclusively in the germ cells of the gonads

(E) It may be used for determination of chromosomal sex and abnormalities of X-chromosome number

68. Which of the following occurs during interphase?

(A) Duplication of the centrosome during a specific phase of interphase

(B) Reformation of the nuclear envelope around individual chromosomes

(C) Disappearance of the nucleolus

(D) Separation of sister kinetochores

(E) Formation of the contractile ring

69. Which of the following events occurs in meiotic prophase I?

(A) DNA synthesis occurs

(B) Paired maternal and paternal chromosomes undergo gene exchange

(C) Maternal and paternal chromosomes are segregated at random

(D) The chromosome content of the cell is halved

(E) Homologous chromosomes are aligned on the metaphase plate of the meiotic spindle

70. Which of the following statements is true of microtubule function during mitosis?

(A) Kinetochore microtubules grow at the kinetochore-attached positive end by the addition of tubulin

(B) Astral microtubules form the part of the spindle apparatus directly related to chromosomal movements

(C) Polar microtubules do not participate in the formation of the mitotic spindle

(D) Spindle microtubules are unaffected by drugs such as colchicine

(E) Anaphase events are independent of microtubular actions

DIRECTIONS: Each numbered question or incomplete statement below is NEGATIVELY phrased. Select the **one best** lettered response.

71. Cell fusion experiments with synchronized cells are carried out with the following results:

1. When an S-phase cell is fused with a G_1 cell, the G_1 cell is driven into S phase.

2. When a G_1 cell is fused with a G_2 cell, the G_1 remains on its own timetable and the G_2 nucleus remains in G_2.

3. When a G_2 cell is fused with an S cell, the S-phase cell proceeds through the cell cycle until it reaches G_2 phase. Both cells then enter M phase in synchrony.

All the following are correct interpretations relating to cell cycle regulation EXCEPT

(A) the S-phase cell contains an activator that causes the G_1 cell to enter S phase prematurely

(B) the S-phase activator is no longer effective when cells complete S phase and enter G_2 phase

(C) mitosis is delayed by incompletely replicated DNA, which may be present during S phase

(D) there is a re-replication block that prevents reduplication of DNA

(E) a factor present in G_2 induces cells to bypass G_1 and S phase and enter G_2

72. All the following statements are true of the cell cycle EXCEPT

(A) the cell cycle consists of chromosomal and cytoplasmic cycles in which DNA and cytoplasmic components are duplicated and divided between the two daughter cells

(B) the cell cycle consists of the following stages in order: interphase, prophase, prometaphase, metaphase, anaphase, and telophase

(C) cytokinesis represents the condensation of chromosomal material with two sister chromatids held together at the centromere

(D) the duration of the cell cycle and its individual phases varies from organ to organ and during development and senescence

(E) normal passage through the cell cycle requires accurate inheritance and duplication of the centrosome

DIRECTIONS: Each group of questions below consists of lettered headings followed by a set of numbered items. For each numbered item select the **one** lettered heading with which it is **most** closely associated. Each lettered heading may be used **once, more than once, or not at all.**

Questions 73–74

Match each description to the most appropriate structure labeled in the electron micrograph below.

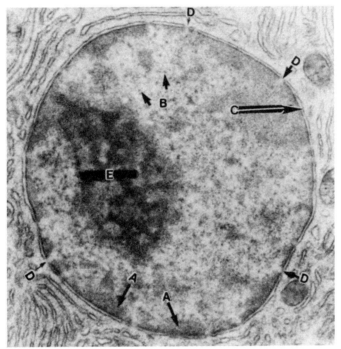

From Fawcett DW: *A Textbook of Histology,* 11/e. WB Saunders, 1986, with permission.

73. A structure visible at the light microscopic level; it generally is believed to represent chromatin that is transcriptionally inactive during interphase

74. A structure that may be labeled using ^{3}H-uridine and autoradiography at the ultrastructural level

Questions 75–77

The figure below summarizes the molecules involved in the cyclin cycle for regulation of entry into the mitotic phase of the cell cycle. Match each description with the best possible lettered answer from the figure. The dark stippling in the oval represents inactivation of that component.

CYCLIN CYCLE

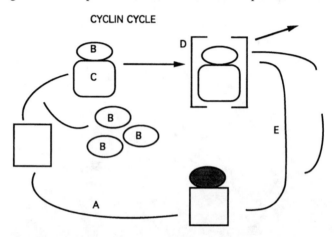

75. This molecule directly phosphorylates lamins

76. This molecule, originally isolated by the study of yeast mutants, subserves a kinase function

77. This event occurs during the mitotic phase of the cell cycle

---

Questions 78–79

Match each mitotic event with the appropriate stage of the cell cycle.

(A) Prophase
(B) Anaphase A
(C) Anaphase B
(D) Telophase
(E) Prometaphase

78. Nucleoli disappear

79. Polar microtubules elongate as the two poles of the spindle move further apart

Questions 80–81

Match each event in meiosis to the appropriate stage

(A) Leptotene
(B) Zygotene
(C) Pachytene
(D) Diplotene
(E) Diakinesis

80. Crossing-over

81. Condensation of chromatids

Cell Biology: Nucleus

Answers

63. The answer is B. *(Junqueira, 8/e, pp 52–53. Ross, 3/e, p 47. Stevens, pp 11–12.)* The nucleolus is a highly organized, heterogeneous structure within the nucleus with distinct regions usually visible by electron microscopy: (1) fibrillar centers, which represent the nucleolar organizer regions where DNA is not being actively transcribed; (2) dense fibrillar components (pars fibrosa) where RNA molecules are being transcribed; and (3) a granular component (pars granulosa) where ribosomes undergo maturation. The nucleolar organizer contains clusters of rRNA genes (DNA). Ribosomal synthesis and much of the ribosomal assembly process occur in the nucleolus, but the maturation of ribosomes requires transport to the cytoplasm. Ribosomal proteins as well as all proteins that function in the nucleus are synthesized in the cytosol and transported into the nucleus. The size and number of nucleoli differ with the metabolic activity of cells.

64. The answer is B. *(Alberts, 3/e, pp 863–866, 917, 934–937. Junqueira, 8/e, pp 54–58.)* The mitotic cell in the micrograph is undergoing cytokinesis, or the cleavage of the cytoplasm to form two cells. Cytokinesis occurs after completion of nuclear division (i.e., nuclear condensation and separation of the chromosomes). It requires the action of the contractile ring composed of actin and myosin. The contractile ring pulls the plasma membrane of the telophase cell into the cleavage furrow. Kinetochore microtubules, which attach the kinetochores to the spindle apparatus, shorten and result in the pulling of chromatids to opposite poles. Growth of polar microtubules results in the separation of the spindle poles. Both of these events occur in anaphase, which is part of mitosis, not cytokinesis. The force for cytokinesis is generated by the action of actin and myosin, which may be inhibited by treatment with antimyosin antibodies in vitro.

65. The answer is A. *(Alberts, 3/e, pp 342–343, 353–354, 384–385, 561–562. Goodman, pp 108–111.)* Nucleosomes are the basic structural packaging unit of chromatin. Chromatin strands that have been treated to unpack the chromatin structure have the appearance of beads on a string in electron micrographs. The beads are formed by a core of histones as an octamer (i.e., two of each of the four nucleosomal histones: H2A, H2B, H3, and H4) plus two turns of DNA. The nucleosome beads plus the DNA between beads (i.e., linker DNA) constitute the nucleosome. There are additional orders of chromosome packing, including nucleosomal packing. Adjacent nucleosomes are packed together through the binding of histone H1 to the nucleosome region in which the DNA helix exits and enters the histone nucleosomal octamer. The transcription of DNA is carried out by RNA polymerases I, II, and III, which are responsible for transcription of different types of genes. The nuclear pores are perforations in the nuclear envelope, each composed of a nuclear pore complex. The nuclear matrix is the intranuclear cytoskeleton and forms the scaffolding for nuclear structures. Chromatids are held together at the centromere.

66. The answer is A. (*Alberts, 3/e, pp 569–570, 800–801. Goodman, p 131. Junqueira, 8/e, p 48. Stevens, pp 11, 21–22. Widnell, pp 76–83.*) Lamins are a subclass of intermediate filaments including three nuclear proteins: lamins A, B, and C. The lamins differ from other intermediate filament proteins in some structural respects, but more importantly in the presence of a nuclear import signal. The lamins form the core of the nuclear lamina, interact with nuclear envelope proteins, and play a role in the maintenance of the shape of the nucleus. Phosphorylation of intermediate filaments leads to disassembly as occurs with the lamins. The disassembly of lamins results in the dissolution of the nuclear envelope as occurs in prometaphase of the cell cycle. Dephosphorylation, not phosphorylation, of the lamins is associated with the reassembly of the nuclear envelope in telophase, where the membrane reforms around the chromosomes. Porins are transmembrane proteins that form pores in the outer membrane of mitochondria and gram-negative bacteria. Chaperonins are cytosolic protein chaperones essential for the proper unfolding of proteins. The heat shock protein (hsp 70) responsible for the unfolding of proteins within the mitochondrial matrix is an excellent example of this class of proteins.

67. The answer is E. (*Junqueira, 8/e, pp 51–53. Moore, Before We are Born, 4/e, pp 119–120. Moore, Developing Human, 5/e, pp 106–107. Ross, 3/e, pp 46–47.*) Females have two X chromosomes, one of maternal and the other of paternal origin. Only one of the X chromosomes is active in the somatic, diploid cells of the female; the other X chromosome remains inactive and is visible in appropriately stained interphase cells as a mass of heterochromatin. Detection of the Barr body (sex chromatin) is an efficient method for the determination of chromosomal sex and abnormalities of X-chromosome number; however, it is not definitive proof of maleness or femaleness. In Turner's syndrome (XO) no Barr bodies would be present. In comparison, "superfemales" (XXX) and persons with Klinefelter's syndrome (XXY) would both possess two inactive X chromosomes, although the sex of these persons would be female and male, respectively, as determined by the presence or absence of the testis-determining Y chromosome. Since only one X chromosome is present in each diploid cell, normal females are mosaic for the X chromosome in that relatively equal clones of X-maternal and X-paternal cells occur.

68. The answer is A. *(Alberts, 3/e, pp 863–866, 916–917. Junqueira, 8/e, pp 54–58.)* Interphase consists of three phases: G_1, S, and G_2. The S phase is the period in which replication of DNA occurs. The G_1 and G_2 phases of the cell cycle represent gap phases in which biosynthetic preparations for DNA synthesis (G_1 phase) and mitosis (G_2 phase) occur, respectively. The M phase follows the G_2 phase. During prophase the nucleolus disappears as the chromosomes condense. At this juncture, the two sister chromatids of each chromosome are held together in the centromere. In prometaphase the nucleolus disappears and the plaquelike kinetochores form on each centromere. In metaphase, the chromosomes are aligned in the equatorial plane by the action of the kinetochores, which rapidly separate at the end of metaphase. Cytokinesis begins during anaphase and is completed during late telophase through the action of the contractile ring.

69. The answer is B. *(Alberts, 3/e, pp 1014–1021. Widnell, pp 66–71.)* Meiosis is the mechanism used by the reproductive organs to generate gametes, cells with the haploid number of chromosomes. DNA synthesis occurs before meiotic prophase I begins and is followed by a G_2 phase. Cells then enter meiotic prophase I. During meiotic prophase I, maternal and paternal chromosomes are precisely paired and recombination occurs in each pair of homologous chromosomes. There are five substages of meiotic prophase I: leptotene, zygotene, pachytene, diplotene, and diakinesis. During metaphase I there is random segregation of maternal and paternal chromosomes. The second meiotic division is responsible for the reduction in the chromosome content of the cell by 50 percent. Homologous chromosomes are aligned on the metaphase plate of the meiotic spindle in metaphase I. In meiotic division II, metaphase consists of daughter chromatids of single homologous chromosomes aligned on a metaphase plate (metaphase II).

70. The answer is A. *(Alberts, 3/e, pp 924–925, 929–933.)* Tubulin is added to the positive end of kinetochore microtubules near their attachment to kinetochores during metaphase; a reversal of this pattern is observed during anaphase. Astral microtubules extend away from the spindle poles and do not form part of the spindle apparatus associated directly with chromosomal movement. The microtubules directly associated with spindles include the polar and kinetochore microtubules. Spindle microtubules are disrupted by treatment with colchicine, which prevents mitosis. Anaphase events are separated into two parts. Anaphase A involves the shortening of kinetochore microtubules, which pulls the chromatids toward opposite poles. In anaphase B, growth of polar microtubules results in the separation of the spindle poles. Both stages of anaphase are dependent on microtubular function.

71. The answer is E. (*Alberts, 3/e, pp 872–873, 878–879.*) The first cell fusion experiment indicates that a factor in the S-phase cell (i.e., an S-phase activator) induces the G_1 cell to enter S phase. The second cell fusion experiment provides evidence that once a cell leaves S and enters G_2 phase, the S-phase activator is inhibited or destroyed. The third experiment suggests that the delay in the G_2 cell is caused by the incomplete replication of DNA in the S-phase cell. Once the S-phase cell completes S and enters the G_2 phase, all DNA is replicated (i.e., incompletely replicated DNA is no longer present). This experiment also demonstrates that there is a re-replication block. For example, the G_2 cell cannot be induced to reenter S phase. There is no evidence that G_2 induces a shortening of preceding phases of the cell cycle.

72. The answer is C. (*Alberts, 3/e, pp 911–918.*) The cell cycle is the process by which cells duplicate nuclear and cytoplasmic contents and distribute them to two new daughter cells. One may consider the cell cycle to be composed of three component cycles: chromosomal, cytoplasmic, and centrosomal. In the chromosomal cycle the genomic content (DNA) is duplicated, the chromosomes condense and are aligned in the metaphase plate, and the chromosomes are subsequently separated into two daughter cells. The cytoplasmic cycle represents the duplication of the cytoplasmic content and the separation of the cytoplasmic components into two daughter cells. In the centrosomal cycle the centrosome, which consists of a pair of centrioles, duplicates and separates during interphase and forms the spindle poles of the mitotic apparatus during metaphase. During cytokinesis the centrosomes return to their prereplicated, interphase state. The cell cycle consists of distinct stages: interphase, prophase, prometaphase, metaphase, anaphase, and telophase. From prophase to telophase is referred to as the *M phase* of the cell cycle in which chromosomal condensation and segregation of duplicated materials to the two daughters cells occur. Cytokinesis represents the cleavage of one parental cell into two daughter cells. In prophase, chromosomes condense and consist of two sister chromatids held together at the centromere. The role of the centrosomes in forming the mitotic spindles establishes the appropriate inheritance and duplication of these structures as essential events in the cell cycle. The duration of the cell cycle varies from organ to organ and during development. Cells such as those in the crypts of the small intestinal epithelium traverse the cell cycle rapidly, while osteoprogenitor cells have a much longer cell cycle. Also, duration of the cell cycle varies during development. Embryonic or early postnatal organs, which are undergoing rapid growth, usually contain cells that quickly traverse the cell cycle. Most of the variation in the time of the cell cycle occurs in the G_1 phase.

73–74. The answers are 73-A, 74-E. *(Junqueira, 8/e, pp 48–52. Ross, 3/e, pp 43–45, 47–49. Stevens, pp 11–13.)* There are two subclassifications of chromatin on a morphological basis. Heterochromatin (A) is visible with the light microscope as condensed basophilic clumps and with the electron microscope as compact, electron-dense material within the nucleus. It is transcriptionally inactive during the interphase stage of the cell cycle, when the genetic material is normally duplicated. Euchromatin (B) is actively transcribed chromatin and is visible only with the use of electron microscopy. Cells with extensive euchromatin are considered metabolically active.

The nucleolus (E) is the site of ribosomal RNA syntehsis. ^{3}H-uridine may be localized in the nucleolus by use of autoradiography and is often used as a marker for RNA synthesis since uridine is preferentially incorporated into RNA. RNA is packaged with ribosomal proteins to form ribosomes.

The nuclear envelope (C) shields the nucleus from the cytoplasm, which allows the sequestration of the genetic material from mechanical cytoplasmic forces. The separate nuclear compartment also allows for separation of the cellular processes of transcription and translation. The nuclear envelope consists of two concentric unit membranes. The outer membrane is continuous with the rough endoplasmic reticulum. Ribosomes on the outer nuclear membrane release proteins into the perinuclear space (cisterna) between the inner and outer nuclear membranes. The nuclear envelope is therefore associated with protein synthesis, and the outer nuclear membrane may be considered part of the rough endoplasmic reticulum with regard to ribosomal function since the perinuclear cisterna is continuous with the cisterna of the rough endoplasmic reticulum. The inner nuclear membrane is associated with a lamina of fibrous proteins including intermediate filament proteins, known as *lamins,* which appear to regulate the assembly and disassembly of the nuclear membrane during mitosis.

Nuclear pores (D) are interruptions in the nuclear envelope that function as aqueous channels for the passage of soluble molecules from the nucleus to the cytoplasm (ribosomes) and from the cytoplasm to the nucleus (nuclear proteins synthesized in the cytoplasm and transported to the nucleus). There are thousands of nuclear pores in the average nuclear envelope. The nuclear pores form from the fusion of the inner and outer membranes at the margin of each pore. The nuclear envelope is highly selective with selection based upon pore size, the presence of nuclear import signals, and receptor recognition of RNAs.

75–77. The answers are 75-D, 76-C, 77-B. *(Alberts, 3/e, pp 885–890. Goodman, pp 170–172.)* The cell cycle is regulated by a number of regulatory proteins. One of the primary regulators of the cell cycle is the p34-cdc kinase (C), which was originally studied in yeast mutants. It is combined with cyclin (cytoplasmic oscillator) to form maturation-promoting factor (MPF; D). The cdc2 kinase has a similar concentration during all phases of the cycle; however, its enzymatic activity increases during the transition from G_2 to M phase as the cyclin (B) levels increase. The degradation of the p34-cdc kinase complex occurs during the M phase (mitosis). Mitosis is controlled because cyclin accumulates during interphase and associates with the cdc2 protein to form pre-MPF, an inactive form of MPF. Then enzymes convert the complex into active MPF, which triggers mitosis, a series of phosphorylations including histones and lamins, and activates enzymes that degrade cyclin. As cyclin is destroyed, MPF disappears and the cyclin-degrading enzymes are inactive. At this point cyclin accumulates again. The start (M/G_1) point is regulated in a similar fashion by cdc2 and a second form of cyclin.

78–79. The answers are 78-A, 79-C. *(Alberts, 3/e, pp 916–917. Junqueira, 8/e, pp 54–55. Widnell, pp 58–64.)* The mitotic phase of the cell cycle consists of prophase, prometaphase, metaphase, anaphase, and telophase. In prophase of mitosis, the nucleoli disappear and the chromosomes condense. When the nuclear envelope breaks down, the cell enters the prometaphase stage. The prometaphase stage is also marked by the maturation of the kinetochores on each centromere for the attachment of the chromosomes to the kinetochore microtubules. In metaphase the chromosomes align along the equatorial (metaphase) plate. The sudden separation of sister kinetochores results in the entrance of cells into anaphase, which is separated into two phases. In anaphase A, the mechanism of action is the shortening of the kinetochore microtubules, which results in the movement of the chromosomes toward the poles. In anaphase B, tubulin is added to the positive ends of the microtubules (kinetochore end of the microtubules), which causes elongation and the movement of the poles further apart. Telophase is marked by the reformation of the nuclear envelope.

80–81. The answers are 80-C, 81-A. *(Alberts, 3/e, pp 1014–1021. Widnell, pp 66–71.)* The first meiotic prophase consists of five substages: leptotene, zygotene, pachytene, diplotene, and diakinesis. Condensation of the chromatids occurs in leptotene. In zygotene the synaptonemal complex begins to form, which initiates the close association between chromosomes known as *synapsis.* The bivalent is formed between the two sets of homologous chromosomes (one set maternal and one set paternal equals a pair of maternal chromatids and a pair of paternal chromatids). The four chromatids form a tetrad (bivalent). Pachytene begins as soon as the synpasis is complete and includes the period of crossing-over. The fully formed synaptonemal complex is present during the pachytene stage. At each point where crossing-over has occurred between two chromatids of the homologous chromosomes, an attachment point known as a *chiasma* forms. The formation of chiasmata and desynapsing (separation of the axes of the synaptonemal complex) occur in the diplotene stage. Diakinesis is an intermediate phase between diplotene and metaphase of the first meiotic division.

Epithelium

DIRECTIONS: Each question below contains five suggested responses. Select the **one best** response to each question.

82. The function of microvilli is

(A) extensive movement of substances over cell surfaces
(B) increase in surface area for absorption
(C) cellular movement
(D) transport of intracellular organelles through the cytoplasm
(E) specialized uptake of macromolecules

83. Pemphigus is a disease in which persons make antibodies to one of their own skin cadherin proteins that is involved in the formation of spot-welds between cells. Which of the following junctional complexes would be most affected in this disease?

(A) Macula adherens
(B) Hemidesmosomes
(C) Zonula occludens
(D) Gap junctions
(E) Zonula adherens

84. The mechanism for tube formation as occurs during development of the neural tube could best be explained by

(A) contraction of microfilament bundles associated with the macula adherens
(B) increased condensation of the transmembrane linkers of the desmosomes
(C) increased numbers of sealing strands in the zonula occludens
(D) condensation of the gap junctions
(E) contraction of tonofilaments associated with the desmosomes

85. In the figures below, A is a transmission electron micrograph and B is a freeze-fracture preparation of a

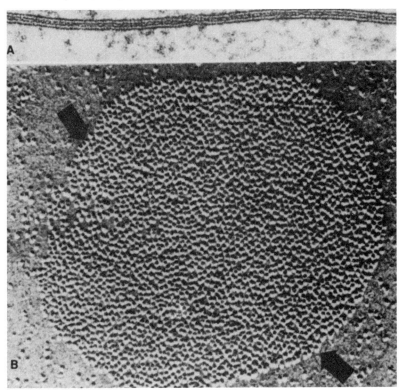

From Fawcett DW: *The Cell,* 2/e, WB Saunders, 1981, with permission. Courtesy of Drs. K. Hama and D. Albertini.

(A) macula adherens
(B) zonula adherens
(C) terminal web
(D) terminal bar
(E) gap junction

DIRECTIONS: Each numbered question or incomplete statement below is NEGATIVELY phrased. Select the **one best** lettered response.

86. All the following statements are true of epithelia EXCEPT

(A) they typically demonstrate an abundance of intercellular substance
(B) they demonstrate strong adhesion between cells
(C) they may be derived from any of the embryonic germ layers
(D) they are capable of performing absorptive functions
(E) they are capable of secretion

87. Which of the following is LEAST characteristic of epithelial polarity?

(A) Mitochondrial distribution
(B) Perinuclear location of the Golgi apparatus
(C) Junctional complexes on basolateral surfaces
(D) Na$^+$,K$^+$-ATPase distribution
(E) Distribution of smooth endoplasmic reticulum in a steroid-secreting cell

88. All the following are responsible for the maintenance of epithelial polarity EXCEPT

(A) tight junctions
(B) targeted delivery
(C) redistribution of proteins in the membrane
(D) sorting of domain-specific proteins in the endoplasmic reticulum
(E) the presence of an intact basement membrane

89. Basement membrane functions include all the following EXCEPT

(A) molecular filtering
(B) selective blocking of the movement of cells
(C) facilitation of repair processes
(D) guiding of cell migration
(E) active ion transport

90. The lamina rara of the basement membrane contains all the following EXCEPT

(A) proteoglycans
(B) adhesive proteins
(C) type IV collagen
(D) fibronectin
(E) laminin

91. The apical and basolateral surfaces of epithelial cells have all the following in common EXCEPT

(A) a capacity for ion transport
(B) presence of membrane receptors
(C) presence of a glycocalyx
(D) a site of exocytosis
(E) a site of endocytosis

92. All the following statements are true of the basal membrane of an epithelial cell EXCEPT

(A) it contains an enzyme that is inhibited by ouabain
(B) it is a major site for hormone receptors
(C) it is a major site for integrins
(D) it is an important site for uptake of nutrients
(E) it is not involved in endocytosis

93. All the following statements are true of dynein EXCEPT

(A) mutations of dynein lead to immotile cilia syndrome
(B) it is found in the inner and outer arms of the microtubule doublets
(C) it is the functional equivalent of myosin found in the microtubules
(D) it provides energy for sliding
(E) it moves vesicles unidirectionally along microtubules

94. Basal folds can be described by all the following statements EXCEPT

(A) they serve to increase surface area
(B) they probably function in absorption
(C) they are often morphologically associated with large numbers of mitochondria
(D) they are found in the distal tubule cells of the kidney
(E) they may be visible at the light microscopic level as basal striations

95. Kartagener's syndrome is characterized by dynein deficiency, which may cause all the following EXCEPT

(A) bronchiectasis
(B) sinusitis
(C) infertility in the male
(D) infertility in the female
(E) immotile stereocilia

96. All the following statements are true of the lateral membrane of an epithelial cell EXCEPT

(A) it contains enzymes for absorption
(B) it contains enzymes for transepithelial transport functions
(C) it may contain enzymes that are also found in the basal membrane
(D) it is the site of intercellular junctions
(E) lateral membranes of adjacent cells form the terminal web

DIRECTIONS: Each group of questions below consists of lettered headings followed by a set of numbered items. For each numbered item select the **one** lettered heading with which it is **most** closely associated. Each lettered heading may be used **once, more than once, or not at all.**

Questions 97–99

Match each structure with the appropriate label in the transmission electron micrograph below.

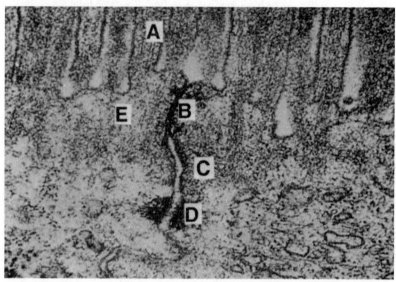

From Erlandsen SL, Magney JE: *Color Atlas of Histology,* CV Mosby, 1992, with permission.

97. Macula adherens

98. Zonula occludens

99. Terminal web

Questions 100–103

Match each description with the appropriate epithelium

- (A) Simple columnar epithelium
- (B) Stratified squamous epithelium
- (C) Transitional epithelium
- (D) Pseudostratified ciliated epithelium
- (E) Simple squamous epithelium
- (F) Stratified columnar epithelium
- (G) Stratified cuboidal epithelium

100. Found in lining of digestive tract; involved in absorption

101. Found in the skin, where it is keratinized; has a protective function

102. Found in the urinary tract; cells vary in shape with degree of stretch

103. Found in blood vessels; structure facilitates transport functions

Questions 104–106

The figure below represents a cross-section of a cilium. Match each numbered item with the appropriate lettered structure.

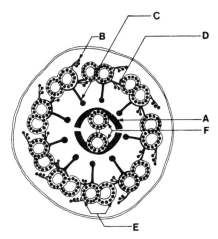

104. Source of ATPase activity

105. Production of bending

106. Inhibition of sliding between doublets

Questions 107–111

Match each numbered description with the correct lettered structure.

(A) Macula adherens
(B) Zonula occludens
(C) Zonula adherens
(D) Gap junction
(E) Hemidesmosome
(F) Focal contacts

107. Important role in the selective-barrier function of epithelia

108. Responsibility for the linkage of the intermediate filament network of cells to the basal lamina

109. Establishment and maintenance of apical versus basolateral polarity of epithelial cells

110. Linkage of actin filaments to the extracellular matrix through interaction of actin-binding proteins with integrins

111. Electrical and chemical coupling between smooth or cardiac muscle cells and between early embryonic cells

Questions 112–114

Match each numbered structure with the appropriate lettered choice.

(A) Triplet arrangement of microtubules
(B) "9 + 2" arrangement of microtubules
(C) Singlet arrangement of microtubules
(D) Microvilli
(E) Minimyosin

112. Centriole

113. Basal body

114. Stereocilia

Epithelium

Answers

82. The answer is B. *(Burkitt, 3/e, p 87. Junqueira, 8/e, pp 66, 68. Ross, 3/e, pp 58–59, 72–73. Stevens, pp 26–27, 32–33.)* Microvilli are apical specializations of the epithelia that form the brush border observed at the light microscopic level. They increase surface area, are relatively uniform in length, and are located on the luminal surface of cells such as small intestinal enterocytes, which are specialized for absorption. Microvilli are supported by a core of microfilaments and are capable of movement, but surfaces with cilia such as those found in the trachea primarily function in the movement of substances such as mucus and the foreign material that adheres along their surfaces. The increased surface area of microvilli facilitates specialized uptake of molecules, which occurs by the processes of pinocytosis, receptor-mediated endocytosis, and phagocytosis, depending on the size and solubility of the molecules. Microtubules facilitate organellar movement within the cytoplasm and cell movement is controlled by interactions between the cytoskeleton and the extracellular matrix.

83. The answer is A. *(Alberts, 3/e, p 956, Goodman, p 188. Isselbacher, 13/e, pp 286–287. Stevens, p 31.)* Pemphigus is a severe skin disease in which patients make antibodies to one of their own skin cadherin proteins. This desmosomal-cadherin molecule interacts with plakoglobin, a constituent of both the macula adherens (desmosome) and the zonula adherens (adhesion belt). Cadherins are Ca^{2+}-dependent transmembrane-linker molecules essential for cell-cell contact. The desmosome (macula adherens) forms the spot-welds between cells and is disrupted in the autoimmune disease pemphigus, resulting in severe blistering and fluid loss.

84. The answer is A. *(Alberts, 3/e, p 1045. Goodman, p 191.)* In the formation of tubular structures from flat sheets, there is contraction of the microfilament bundles associated with the adhesion belt junctions (macula adherens). This occurs in neural tube formation, which involves the conversion of the neural plate into the neural tube. In the apical part of the cells, the apical actin filaments contract, narrowing the cells at their apical ends. The position of the zonula adherens, forming a contractile ring around the circumference of the cell, coupled with the contractile nature of the actin microfilament bundles is ideal for regulating morphogenetic changes.

85. The answer is E. *(Burkitt, 3/e, p 85. Erlandsen, p 20. Junqueira, 8/e, pp 64–65, 67.)* The transmission and freeze-fracture electron micrographs illustrate the structure of a nexus, or gap junction. Gap junctions are composed of multipass transmembrane proteins that form connexons. When the connexons of adjacent cells are in alignment, a pore of about 1.5 nm is open and there is continuity between the interior of the two cells. It is postulated that the channel is formed in a fashion similar to the way a pore is formed by the multipass transmembrane proteins associated with the acetylcholine receptor. The gap junction maintains electrical or chemical coupling or both between cells. In the freeze-fracture micrograph, the intramembranous particles (connexons) are seen in circular arrangements on the P face of the membrane.

86. The answer is A. *(Junqueira, 8/e, p 60. Ross, 3/e, pp 58–60. Stevens, pp 26–27.)* Epithelial, connective, muscle, and nervous tissues are the histologic tissues of the body. Epithelium typically demonstrates a paucity of intercellular substance. Strong adhesion between cells is exemplified by the presence of junctional complexes between adjacent cells. Epithelial cells perform diverse functions including absorption (the small intestinal epithelium) and secretion (glandular epithelium of endocrine or exocrine glands). Epithelia may be derived from ectoderm (epidermis), endoderm (small intestinal epithelium), or mesoderm (endothelium of blood vessels).

87. The answer is E. *(Alberts, 3/e, pp 631–632. Junqueira, 8/e, pp 73–74. Widnell, pp 134–135.)* Cellular polarity is one of the primary characteristics of epithelial cells. Most of the organelles of epithelial cells show polarization, e.g., the perinuclear location of the Golgi and the localization of mitochondria in areas of energy utilization. The apical region of epithelial cells may be covered with specialized structures such as the microvilli, stereocilia, or cilia. The lateral surfaces exhibit junctional complexes that are specialized for attachment between adjacent cells. Basolateral membranes possess enzymes such as Na^+, K^+-ATPase that are not found in the apical membrane. Membrane proteins and lipids also show polarity in the sense that apical and basolateral membranes differ in their content. Smooth endoplasmic reticulum fills the entire cytoplasm of steroid-secreting cells with little polarity.

88. The answer is D. *(Alberts, 3/e, pp 631–632. Junqueira, 8/e, pp 73–74. Widnell, pp 134–135.)* There are a number of factors responsible for the maintenance of epithelial polarity. Tight junctions prevent intermixing and exchange of lipids and proteins between the apical and basolateral domains. An intact basement membrane also helps maintain the polarity of epithelial cells. Proteins destined for transport to apical and basolateral domains follow similar pathways through the endoplasmic reticulum. Sorting of apical versus basolateral domain proteins occurs in the *trans*-Golgi, at which point there is targeted delivery. Redirection of missorted proteins also occurs using endocytosis from the incorrect membrane domain followed by endosomal trafficking to the correct domain. This process effectively represents transcytosis.

89. The answer is E. *(Alberts, 3/e, pp 992–993. Goodman, pp 198–199. Junqueira, 8/e, pp 60–61, 63. Ross, 3/e, pp 61–66.)* The basement membrane is found underlying, enveloping, or interposed between epithelial cells. It serves a number of functions. Epithelial cells require a basement membrane as a structural support. In most epithelia the basement membrane serves to prevent penetration from the underlying lamina propria into the epithelium. Basement membranes are a pathway for migrating cells during development (e.g., glandular morphogenesis) and repair processes (e.g., healing of skin wounds). In the kidney, the basement membrane of the renal glomerulus forms a selective barrier for the filtration of the plasma. Active ion transport is a characteristic of the epithelia that are positioned on the basement membrane, not of the basement membrane.

90. The answer is C. *(Alberts, 3/e, pp 989–991. Goodman, pp 198–199. Junqueira, 8/e, pp 60–61, 63. Ross, 3/e, pp 61–66. Widnell, pp 148–149.)* The basement membrane provides support for epithelia and is synthesized by both epithelial cells and the subjacent connective tissue. Basement membranes primarily contain type IV collagen, heparan sulfate proteoglycans, laminin, and entactin. Although these elements are the primary constituents, basement membrane composition, appearance, and function differ from tissue to tissue and between different epithelia. The basement membrane may be visualized at the light microscopic level, particularly with the periodic acid–Schiff reagent. This method does not separate epithelially derived (basal lamina) and connective tissue–derived (reticular lamina) components of the basement membrane, so that the basement membrane appears uniform from tissue to tissue. At the ultrastructural (electron microscopic) level, basement membranes are composed of one or two electron-lucent areas (laminae rarae), which contain fibronectin, laminin, proteoglycans, and adhesive proteins. Deep to the lamina rara is the lamina densa with its electron-dense staining type IV collagen. The third layer is the reticular layer. This reticular lamina is composed of collagen fibrils formed by the connective tissue below the epithelium. The reticular lamina is found in some, but not all epithelia. Integrins (i.e., laminin and fibronectin receptors) are important in the maintenance of epithelial adhesion to the basement membrane and the adhesion of connective tissue cells to the opposite surface of the basement membrane. Fibronectin is found primarily on the connective tissue side of the basement membrane, while laminin is found toward the epithelial side.

91. The answer is C. *(Alberts, 3/e, pp 502–503, 631–633. Junqueira, 8/e, pp 66–68, 73–74. Widnell, p 134.)* The apical surfacce of epithelia differs from the basolateral membranes in its content of proteins. The apical membrane of the small intestinal enterocyte, for example, is covered with a brush border containing the enzymes involved in digestive processes, such as the disaccharidases maltase and sucrase. The apical surface of kidney tubules contains alkaline phosphatase, while endothelial cells of blood vessels are specialized with the presence of angiotensin-converting enzyme, which converts angiotensin I to angiotensin II. The apical epithelial surface is covered by a glycocalyx. Extending from the apical membrane are oligosaccharides and proteoglycans, including sialic acid residues, which present a negative charge to the luminal surface. The apical surface is the most common site of endocytosis in which clathrin-coated pits invaginate to form clathrin-coated vesicles in a receptor-mediated process; however, endocytosis does occur from the basolateral membranes. An example of this is the uptake of IgA by secretory epithelial cells using receptor-mediated endocytosis. Exocytosis occurs frequently at the apical surface, but may also occur at the basolateral surfaces in cells, such as the small intestinal enterocyte, that release chylomicra at their basolateral surfaces. The enterocyte also exhibits endocytosis of IgA from lamina propria plasma cells followed by combination with secretory component, and subsequent release from the cell at the apical surface by exocytosis. This combined endocytosis at one surface coupled with exocytosis at the other surface is known as *transcytosis*. Membrane receptors such as growth factor and hormone receptors are frequently located on the apical surface, but in some instances are localized basolaterally (e.g., histamine receptors on parietal cells in the stomach). Both apical and basolateral surfaces are capable of ion transport.

92. The answer is E. *(Junqueira, 8/e, pp 73–74, 76–77. Widnell, pp 133–134.)* The basal membrane differs primarily from the apical membrane; there are many similarities between the basal and lateral membranes of epithelial cells. The basal membrane is the major site of hormonal and neurotransmitter receptors. For example, in the parietal cell in the fundic glands of the stomach, the basal membrane possesses histamine, acetylcholine, and gastrin receptors. Integrins are extracellular matrix receptors such as the fibronectin and laminin receptor. They are located on the basal membrane and interact with specific components of the extracellular matrix. This is also the location of epithelial synthesis of extracellular matrix components such as fibronectin, laminin, and type IV collagen. Nutrients are taken up by the basal surface of an epithelial cell. The basal membrane participates in endocytosis during processes such as the uptake of immunoglobulins from the basal surface of small intestinal enterocytes. The immunoglobulins are transported across the cell and released by exocytosis at the apical surface (transcytosis) after addition of secretory component by the enterocyte. Basolateral membranes contain Na⁺, K⁺-ATPase, an enzyme that is inhibited by ouabain. This inhibitor is used to label Na⁺, K⁺-ATPase in the membrane in experimental studies.

93. The answer is E. *(Alberts, 3/e, pp 813–814, 817–818. Goodman, p 91. Ross, 3/e, pp 74–75. Widnell, p 186.)* Dynein is an ATPase located in the inner and outer arms of the microtubule doublets and provides the energy for sliding through the breakdown of ATP. Mutations of dynein lead to immotile cilia in Kartagener's (immotile cilia) syndrome. Dynein contains two or three globular heads with ATPase activity. The heads drive the sliding of microtubules in a cilium in a fashion similar to the mechanism used by myosin heads to walk along actin filaments. Nexin is essential in the process of ciliary bending. It holds adjacent doublets in the cilium and converts microtubule sliding into doublet bending. Kinesin, not dynein, moves vesicles unidirectionally along microtubules.

94. The answer is B. *(Ross, 3/e, pp 75–76. Stevens, p 33.)* Basal folds are modifications of the basal region of the cell. These deep infoldings of the basal plasma membrane increase surface area and compartmentalize numerous mitochondria that provide energy for ionic and water transport. Distal tubule cells of the kidney and striated duct cells of the submandibular glands possess predominant basal infoldings, which are observed at the light microscopic level as basal striations.

95. The answer is E. *(Junqueira, 8/e, pp 42–43, 47, 327. Ross, 3/e, pp 74–75. Stevens, p 35. Widnell, p 186.)* In *immotile cilia syndrome* the outer dynein arms may be absent and microtubular arrangements are abnormal. The result is a failure of normal ciliary action. Chronic bronchial and sinus infections are common occurrences in these patients because the cilia are unable to remove foreign material from the bronchi and sinuses. Infertility in the male may be due to absence of normal ciliary proteins in the flagella of the spermatozoa. Infertility in the female may be related to problems in movement of the ovum through the oviduct. Stereocilia are immotile; they are more similar to microvilli than cilia and are not involved in the movement of sperm despite their location in the epididymis and vas deferens. Many of the patients diagnosed with immotile cilia syndrome are observed to have a lateral transposition of the major organs of the body (situs inversus). Normal ciliary action may be required for normal positioning of organs during development.

96. The answer is E. *(Junqueira, 8/e, pp 70–71, 76–77. Widnell, pp 133–134.)* The lateral membrane of epithelial cells has a distinct polarity compared with the apical surface. It shows similarities to the apical and basal membranes. The lateral membrane contains enzymes for absorption and transepithelial transport functions as does the apical membrane. Similarities to the basal membrane include the presence of Na⁺, K⁺–ATPase, which is usually described as an enzyme specific for the basolateral membrane. Na⁺K⁺–ATPase is responsible for the transport of sodium out of the cell and potassium into the cell. Intercellular junctions, which are visualized at the light microscopic level as the terminal bar, are found along the lateral surfaces of epithelial cells. The terminal web is the network of actin and other filamentous proteins (myosin, cytokeratin, and spectrin/fodrin) associated with the apical surface.

97–99. The answers are 97-D, 98-B, 99-E. *(Burkitt, 3/e, pp 82–85. Erlandsen, pp 9, 14, 18, 20. Junqueira, 8/e, pp 62–68.)* The transmission electron micrograph illustrates a junctional complex between two enterocytes in the small intestine. Label A represents the microvilli, which constitute the brush border. The brush border is covered by the glycocalyx and contains enzymes involved in the degradation of food in the lumen of the small intestine. The structure labeled B is the zonula occludens, which provides a tight seal between the epithelial cells. Label C marks the zonula adherens, which interacts with components of the terminal web (label E). The macula adherens (desmosome) is labeled D.

100–103. The answers are 100-A, 101-B, 102-C, 103-E. *(Junqueira, 8/e, pp 60, 62, 68–73. Ross, 3/e, pp 58–60. Stevens, p 27.)* Epithelia perform a multitude of functions. Simple columnar epithelium is involved in absorption or secretion or both and forms the lining of the digestive tract. The stratified squamous epithelium is composed of several layers of cells, which begin with a basal layer of germinative cells where new cells are born and become progressively flattened toward the surface, or lumen. Stratified squamous epithelium provides a protective function and is found in sites such as the skin, where it is keratinized, and in the esophagus and anus, where it is nonkeratinized. The cells of the transitional epithelium increase in size toward the surface, or lumen. This is in contrast to the progressively flattened cells of the stratified squamous epithelium. The cells of the transitional epithelium vary in shape depending upon the degree of stretch of the wall. Transitional epithelium is found lining the urinary bladder, ureters, and specific portions of the urethrae. In the distended bladder, the stratified nature of the epithelium is difficult to discern, but the multilayered appearance can be observed definitively in the empty urinary bladder. The pseudostratified ciliated epithelium is located in the trachea and throughout the respiratory system and in the male genitalia. As the name implies, the pseudostratified epithelium is not actually stratified. Nuclei at different levels present the appearance of stratification, but all cells reach the basal lamina. A simple squamous epithelium lines blood vessels (endothelium) and mesenteries (mesothelium), and its structure facilitates transport functions. Stratified cuboidal epithelium and stratified columnar epithelium line the sweat ducts and the excretory ducts of the parotid gland, respectively.

104–106. The answers are 104-B, 105-B, 106-D. *(Alberts, 3/e, pp 803– 806, 816–817. Burkitt, 3/e, p 86. Junqueira, 8/e, pp 42–43, 68. Ross, 3/e, pp 74– 76. Stevens, pp 34–35. Widnell, p 185.)* Cilia and the structurally similar flagella produce wavelike bending movements for propulsion of materials over the surface (e.g., movement of mucus in the tracheal epithelium) or cellular movement (e.g., that of sperm). The arrangement of the ciliary axoneme is described as a "9 + 2" structure that consists of nine outer doublets (E) of complete "A" and incomplete "B" tubules that surround a central pair of complete tubules (F). The dynein arms (B) project from the nine doublets and produce interaction between the doublets, which causes bending. Dynein is an ATPase that provides the energy for bending. The nexin links (D) hold neighboring doublets together and inhibit sliding between doublets. The radical spokes (C) extend from the doublets toward the central pair and are involved in regulation of the ciliary beat. The inner or central sheath (A) surrounds the central doublet.

107–111. The answers are 107-B, 108-E, 109-B, 110-F, 111-D. *(Alberts, 3/e, pp 950–961. Burkitt, 3/e, pp 82–85. Junqueira, 8/e, pp 62–66. Ross, 3/e, pp 26, 66–72. Stevens, pp 28–32. Widnell, pp 137–142.)* The lateral membranes between adjoining cells form intercellular junctions. These junctions can be classified into three general types: occluding junctions, anchoring junctions, and communicating junctions. Each of these classifications can be subdivided into additional types of junctional complexes.

The occluding junctions consist of one main type of lateral membrane modification: the zonula occludens (also known as the *tight junction*). The zonula occludens consists of the closely apposed membranes of the adjoining cells with intramembranous sealing strands that occlude the intermembranous space. The number of sealing strands is believed to be directly proportional to the tightness of the junction. The tight junction serves two very important functions. First, it prevents the passage of luminal substances between the adjacent cells. Second, it maintains apical versus basolateral polarity. The presence of the zonula occludens prevents the lateral diffusion and the homogeneous distribution of membrane proteins. When cells are separated and the zonulae occludentes broken down, apical and basolateral membrane domains are no longer maintained and many membrane proteins are uniformly distributed in the membrane. In addition to these two roles, the tight junctions prevent the back-diffusion of actively transported substances that enter the intercellular space. The zonula occludens therefore plays an important role in the selective-barrier function of the epithelium.

The anchoring junctions are represented by the zonula adherens and focal contacts, which connect actin with actin and actin with extracellular matrix, respectively. The zonula adherens is also known as the *intermediate junction* and involves interaction with actin filaments from the terminal web. It is somewhat similar to the fascia adherens between cardiomyocytes. In the presence of calcium, actin-binding proteins (e.g., α-actinin or vinculin) bind to a cadherin, a transmembrane glycoprotein that mediates cell adhesion. The zonula adherens transmits motile forces across epithelial sheets and serves a major function in the folding of epithelia, such as during neural tube formation in the embryo. Focal contacts represent connection between the actin filaments of the cell and the extracellular matrix. In this case, talin, α-actinin, or vinculin (actin-binding proteins) bind to the integrins (e.g., fibronectin receptor or laminin receptor).

The desmosome (also known as the *macula adherens*) and the hemidesmosome link intermediate filaments in one cell with the intermediate filaments in adjacent cells and with the basal lamina, respectively. Desmosomes are very well developed in epithelia of organs, such as the skin, that are under tensile and shearing forces. The desmosome acts as a "rivet" or "spot weld" between cells. Unlike the zonula occludens and the zonula adherens, the macula adherens is not continuous around the circumference of the cell. The desmosome is associated with tonofilaments that insert into intracellular plaques composed primarily of proteins called *desmoplakins*. The tonofilaments are intermediate filaments composed of keratin in most epithelial cells, but they consist of desmin filaments in cardiomyocytes and vimentin in astrocytes.

The hemidesmosome, which was once thought to be half a desmosome, is now known to have some distinct molecular components compared with the desmosome. The hemidesmosome interacts with the extracellular matrix molecules within the basal lamina through anchoring fibrils. The tonofilaments do not loop through or form lateral attachments to the desmoplakins but terminate in the plaque. The hemidesmosomes combined with the desmosomes act to distribute tensile forces through the epithelial sheet and the supporting connective tissues.

The nexus, or gap junction, is classified as a communicating junction. It allows direct, selective communication in the form of diffusible molecules with a cut-off size of between 1 and 1.5 kD. The gap junction is extremely important in smooth and cardiac muscle for the actions of peristalsis and cardiac contraction. Gap junctions are also found in large numbers in development, particularly between cells in the eight-cell stage of the early embryo. As cells differentiate, the gap junctions become uncoupled.

112–114. The answers are 112-A, 113-A, 114-D. *(Junqueira, 8/e, pp 42– 43, 68. Ross, 3/e, pp 37–39. Stevens, pp 32–34.)* Microtubules are found in different structural patterns within the cell. The centriole consists of nine microtubule triplets arranged together by linking proteins to form a cartwheel arrangement. The basal body is a centriole associated with the ciliary axoneme. It too has a nine-triplet arrangement of microtubules. Cytoplasmic microtubules are found in the singlet form and undergo constant association and dissociation of tubulin with their plus ends and minus ends, respectively. The axoneme has the classic "9 + 2" arrangement of microtubules. Stereocilia are large, modified microvilli.

Connective Tissue

DIRECTIONS: Each question below contains five suggested responses. Select the **one best** response to each question.

115. In Marfan's syndrome, there are mutations in the fibrillin gene resulting in abnormal structure. Which organ would you expect to be most affected?

(A) Middle cerebral artery
(B) Basilar artery
(C) Aorta
(D) Lymphatic vessels
(E) Superior vena cava

116. The extracellular matrix and the cytoskeleton communicate across the cell membrane through

(A) proteoglycans
(B) integrins
(C) cadherins
(D) intermediate filaments
(E) microtubules

117. In Alport's syndrome there is a defect in the $\alpha 5$ chain of type IV collagen. One would expect to see which of the following symptoms?

(A) Abnormal bone formation
(B) Hematuria
(C) Abnormal hyaline cartilage
(D) Skin abnormalities
(E) Ruptured intervertebral disks

118. The function of fibronectin in the extracellular matrix is to

(A) provide structural support
(B) bind signaling molecules
(C) allow passage of nutrients and waste products
(D) provide elasticity
(E) provide an adhesive molecule to facilitate cell attachment

119. Plasma cells are derived from

(A) neural crest
(B) B lymphocytes
(C) T lymphocytes
(D) monocytes
(E) neutrophils

120. Desmosine and isodesmosine are amino acids unique to elastic fibers. They confer elasticity through

(A) cross-linking of microfibrils
(B) cross-linking of tropoelastin
(C) binding of proteoglycans because of the hydrophilic nature of desmosine and isodesmosine
(D) their affinity for elastase, which results in the destabilization of elastic fibers in situ
(E) electrostatic interactions of type IV collagen and elastic fibers

121. Hydroxylation of proline and of lysine

(A) occurs before incorporation into the polypeptide chains
(B) occurs primarily in the Golgi apparatus
(C) occurs in coated vesicles during transport between the Golgi and the plasma membrane
(D) stabilizes the triple helix and facilitates cross-linking, respectively
(E) occurs rarely in collagen synthesis

122. Which of the following statements is true of synthesis and degradation of type I collagen?

(A) Tropocollagen is released from the cell
(B) Lysyl oxidase is functional in the Golgi apparatus
(C) Procollagen peptidases are functional in the coated vesicles
(D) Collagenases secreted by fibroblasts cleave tropocollagen in the extracellular matrix
(E) The registration peptides are removed from the synthesized pro-α-chains in the rough endoplasmic reticulum (RER)

123. The principal proteoglycan with which collagen type IV interacts is

(A) fibronectin
(B) laminin
(C) entactin
(D) heparan sulfate
(E) dermatan sulfate

124. The extracellular matrix molecules fibronectin, laminin, entactin (nidogen), and tenascin have in common

(A) distribution only in the basement membranes of developing tissues and organs
(B) the classification *glycoproteins*
(C) a collagen-binding region that contains the tripeptide sequence RGD (Arg-Gly-Asp)
(D) a circulating plasma form
(E) their function as integrins

125. Ehlers-Danlos syndrome occurs in several forms. In type IV disease there is a defect in type III collagen synthesis. Which of the following symptoms would be most expected in a patient with this disorder?

(A) Rupture of the intestinal or aortic walls
(B) Hyperextensibility of the integument
(C) Hypermobility of diarthrodial joints
(D) Increased degradation of proteoglycans in articular cartilages
(E) Imperfections in dentin formation

126. Which of the following statements regarding brown adipose tissue is correct?

(A) It produces heat through the uncoupling of the electron transport chain from oxidative phosphorylation
(B) It is directly innervated by the parasympathetic nervous system
(C) It is poorly vascularized
(D) It functions in unilocular energy storage
(E) It is involved in the shivering-initiated mobilization of lipid within adipose tissue

127. Which of the following is a similarity between reticular and type I collagen fibers?

(A) Thickness
(B) Visibility in special preparations such as silver stains
(C) Concentration of glycoproteins
(D) Distribution in organs
(E) Fibrillar composition with cross-banding

DIRECTIONS: Each numbered question or incomplete statement below is NEGATIVELY phrased. Select the **one best** lettered response.

128. All the following statements are true of connective tissue EXCEPT

(A) it separates cells, tissues, or organs from the surrounding environment
(B) it provides support
(C) it functions in defense mechanisms
(D) it mediates cell migration
(E) it consists of extracellular matrix formed solely by connective tissue cells

129. The integrins appear to be large and diverse families of membrane proteins. All the following statements are true of integrins EXCEPT

(A) fibronectin receptor and laminin receptor are two well-studied examples
(B) most integrins appear to recognize an RGD (Arg-Gly-Asp) cell-binding sequence
(C) deficiency in platelet integrins may result in excessive bleeding
(D) integrins are integral membrane proteoglycans that serve as transmembrane linkers
(E) fibronectin receptor mediates interactions between the cytoskeleton and extracellular matrix molecules

130. The highly charged and hydrophilic nature of proteoglycans (e.g., hyaluronic acid) plays an important role in all the following EXCEPT

(A) the retention of anions
(B) the facilitation of cell migration
(C) the filtering function of basement membranes
(D) the high viscosity of the intercellular substance, which hinders the penetration of antigens such as bacteria
(E) the lubricating function of the synovial fluid

131. Tumor metastasis involves all the following EXCEPT

(A) release of tumor cells from adhesion to extracellular matrix
(B) dissolution of basement membrane in the source organ
(C) uniformity of tumor cells in regard to metastatic potential
(D) inhibition of metastasis by antibodies to integrins or by collagenases
(E) cell-cell recognition at the site of the new metastasis

132. Scurvy is characterized by all the following EXCEPT

(A) vitamin C deficiency
(B) poor growth of bone and healing of wounds and fractures
(C) defective pro-α-chains that are unable to form a triple helix
(D) excessive denaturation and proteolytic degradation of collagen
(E) increased secretion and assembly of mature collagen molecules

133. Elastin *differs* from collagen in all the following characteristics EXCEPT

(A) lower levels of lysine and proline
(B) the presence of cross-links
(C) an absence of glycosylation
(D) the secretion of a tropo-molecule
(E) synthesis by smooth muscle cells

134. Degradation of the extracellular matrix is accomplished by all the following EXCEPT

(A) collagenases
(B) urokinase-type plasminogen activator
(C) plasmin
(D) serpins
(E) metalloproteases

135. All the following contribute to the tensile strength of collagen EXCEPT

(A) absence of lysine cross-links
(B) the triple helical arrangement of collagen
(C) interactions with the fibril-associated collagens
(D) intramolecular and intermolecular cross-links
(E) tissue organization of collagen

136. Functions of laminin in the basement membrane include all the following EXCEPT

(A) binding to entactin
(B) filtration in the basement membrane
(C) binding to type IV collagen
(D) binding to heparan sulfate
(E) adhesion of epithelial cells to the basement membrane

137. Proteoglycans in the extracellular matrix have a number of functions, including the ability to bind growth factors and cytokines. The effects of proteoglycans on growth factors include all the following EXCEPT

(A) immobilization of the growth factor to restrict its area of activity
(B) protection from proteases in the extracellular matrix
(C) concentration of the growth factor
(D) sterical blockade of the growth factor
(E) direct proteolysis by the proteoglycan

138. All the following characteristics may be ascribed to macrophages EXCEPT

(A) they secrete cytokines
(B) they respond to cytokines
(C) they phagocytose bacteria
(D) they are the first cells to function extravascularly in acute inflammation
(E) they are attracted to an injury site by chemoattractants

139. Mast cells contain and secrete all the following EXCEPT

(A) slow-reacting substance of anaphylaxis (SRS-A)
(B) IgE
(C) heparin
(D) histamine
(E) eosinophil chemotactic factor of anaphylaxis (ECF-A)

140. Systemic mastocytosis would be expected to produce all the following symptoms EXCEPT

(A) gastritis
(B) peptic ulcer
(C) decreased vascular permeability
(D) fibrosis of specific organs
(E) urticaria

DIRECTIONS: Each group of questions below consists of lettered headings followed by a set of numbered items. For each numbered item select the **one** lettered heading with which it is **most** closely associated. Each lettered heading may be used **once, more than once, or not at all.**

Questions 141–143

Wound healing in the skin is mediated by various cytokines and growth factors and results in a series of repair steps. In regard to wound healing, match each structure or function to the correct molecules.

(A) Type III collagen
(B) Platelet-derived growth factor
(C) Integrins on the surface of platelets
(D) Fibronectin
(E) Dense collagen deposition
(F) Type IV collagen
(G) Type II collagen
(H) Fibrin

141. Clot

142. Scar

143. Regulation of proliferation of fibroblasts

Questions 144–148

Match each structure or function with the correct type of collagen.

(A) Collagen type I
(B) Collagen type II
(C) Collagen type III
(D) Collagen type IV
(E) Collagen type V
(F) Collagen type IX

144. Reticular fibers in lymphoid organs

145. Bone

146. Hyaline and elastic cartilage

147. Fibril-associated collagen with interrupted triple helices (FACIT)

148. Extensive interaction with heparan sulfate, which forms the filtration barrier

Questions 149–153

Match each description with the correct cell type.

(A) Fibroblast
(B) Mast cell
(C) T lymphocyte
(D) Macrophage
(E) Plasma cell
(F) Neutrophil
(G) Basophil
(H) Eosinophil
(I) B lymphocyte

149. Active in phagocytosis of antigen-antibody complex; otherwise not considered a major phagocyte

150. Metachromatic cells near luminal surface of small intestinal epithelium

151. Accumulation of these dead cells forms pus

152. Sole source of blood histamine

153. Synthesis of IgE

Connective Tissue

Answers

115. The answer is C. *(Alberts, 3/e, p 986. Isselbacher, 13/e, pp 2115–2116.)* Marfan's syndrome is an autosomal dominant disease in which persons develop abnormal elastic tissue. Malformations include cardiovascular (valve problems as well as aortic aneurysm), skeletal (abnormal height, severe chest deformities) and ocular systems. The molecular basis of the disease is a mutation in the fibrillin gene. The aorta is the most affected organ because of the extensive elastin in the wall, and dissecting aortic aneurysms are common in these patients. The lens is also often affected in patients with Marfan's syndrome. The result is the dislocation of the lens because of loss of elasticity in the suspensory ligament.

116. The answer is B. *(Alberts, 3/e, p 997. Goodman, pp 205–206.)* The integrins are membrane receptors for extracellular matrix components that have the structure of transmembrane heterodimers. The best examples are fibronectin receptor and laminin receptor. The receptor structure includes an intracytosolic portion, which binds to the actin cytoskeleton through the attachment proteins talin or actinin. The extracellular portion has specificity for extracellular matrix molecules. The N-cadherins function as transmembrane glycoproteins involved in the formation of parts of the intercellular junctional complexes.

117. The answer is B. *(Isselbacher, 13/e, p 2117.)* The basement membrane is composed primarily of type IV collagen, heparan sulfate proteoglycan, laminin, and entactin. Basement membranes are usually composed of an electron-lucent layer (lamina rara) closest to the epithelial layer and an electron-dense layer (lamina densa) below the lamina rara. Type IV collagen is found in the lamina densa of the basement membrane. Alport's is an X-linked syndrome in which there is an absence of the α5(IV) chain, resulting in thickening of the basement membrane with splitting of the lamina densa. The disease results in hematuria from the loss of the normal filtering properties of the glomerular basement membrane, leading to nephritis and eventually renal failure.

118. The answer is E. *(Alberts, 3/e, pp 986–987. Goodman, p 198. Junqueira, 8/e, pp 90–91. Stevens, p 46.)* Fibronectin is an adhesive glycoprotein that is important for cellular function in the adult and for cell migration during development. Neural crest and other cells appear to be guided along fibro-nectin-coated pathways in the embryo. Fibronectin is found in three forms: a plasma form, which is involved in blood clotting; a cell surface form, which binds to the cell surface transiently; and a matrix form, which is fibrillar in arrangement. Fibronectin contains a cell-binding domain (RGD sequence), a collagen-binding domain, and a heparin-binding domain. Type IV collagen is responsible for providing support. Elastin is responsible for the elasticity of structures like the pinna of the ear and the wall of the aorta. Proteoglycans are responsible for the passage of molecules and the binding of growth factors and other signaling molecules in the extracellular matrix.

119. The answer is B. *(Junqueira, 8/e, pp 107–110. Male, 2/e, p 3. Roitt, 3/e, pp 2.10–2.12. Ross, 3/e, p 114. Stevens, p 76.)* Plasma cells are derived from B lymphocytes. After stimulation with antigen, the B lymphocyte divides to form a memory B cell and a plasma cell. The plasma cell does not enter the circulation under normal conditions, but remains in the tissues, ful-filling its function of producing antibodies. All immunoglobulins are pro-duced and released by plasma cells.

120. The answer is B. *(Junqueira, 8/e, pp 100–101. Stevens, p 45.)* The two amino acids desmosine and isodesmosine are unique to elastic fibers. They are responsible for the cross-linking of tropoelastin through the lysine residues. Lysyl oxidase is the enzyme that catalyzes this reaction. The mi-crofibrils composed of fibrillin are glycoproteinaceous and facilitate forma-tion of the elastin molecules but are not influenced by lysyl oxidase. Elastase is a serine protease specific for the degradation of elastin. Interactions occur between type III collagen and elastic fibers. The collagen may serve to limit the stretch of the elastic components.

121. The answer is D. *(Alberts, 3/e, pp 980–982. Junqueira, 8/e, pp 94–97.)* Prolyl and lysyl oxidase are the two enzymes that carry out hydroxyla-tion of proline and lysine. The process is both co- and posttranslational and therefore occurs during or more often after the amino acids are inserted into nascent collagen polypeptide chains in the RER. These two amino acids are characteristic of collagen. Hydroxyproline, which constitutes 10 percent of collagen, is often used to determine the collagen content of various tissues. Hydroxylation of proline stabilizes the triple helix through interchain hydro-gen bonds, and hydroxylation of lysine is critical for the cross-linking stage of collagen assembly.

122. The answer is D. *(Alberts, 3/e, p 981. Junqueira, 8/e, pp 94–97. Ste-vens, p 44.)* Collagen is synthesized as pro-α-chains, which are assembled into procollagen molecules (triple helix) in the rough endoplasmic reticulum. Procollagen is subsequently transported in transfer vesicles to the Golgi for packaging into secretory vesicles. Transport of secretory vesicles is an energy-and microtubule-dependent process. Outside of the cell, N-terminal and C-terminal specific procollagen peptidases cleave the nonhelical registration peptides, which results in the formation of tropocollagen. Tropocollagen spontaneously assembles in a staggered array to form collagen fibrils. Lysyl oxidase is an extracellular enzyme responsible for the formation of covalent cross-links between tropocollagen molecules. Fibrils form collagen fibers under the influence of other extracellular matrix constituents, such as proteoglycans and glycoproteins. Collagenases specifically cleave tropocollagen in the extracellular matrix.

123. The answer is D. *(Alberts, 3/e, pp 972–978. Junqueira, 8/e, pp 88–91. Stevens, pp 42–43, 46–47.)* Proteoglycans are large molecules that maintain hydration space in the extracellular matrix. They are composed of glycosaminoglycans such as chondroitin sulfate, dermatan sulfate, heparan sulfate, heparin, and keratan sulfate. All of these particular glycosaminoglycans are covalently linked to core proteins to form proteoglycans. The notable omission from this list is hyaluronic acid, which forms the core of proteoglycan aggregates produced by the interaction between proteoglycan subunits and hyaluronic acid. Dermatan sulfate is found predominantly in the skin, blood vessels, and heart. Heparan sulfate is primarily associated with type IV collagen in basal laminae. Fibronectin is a fibrillar protein, while laminin and entactin are structural glycoproteins found in the extracellular matrix.

124. The answer is B. *(Stevens, pp 46–47.)* The extracellular matrix (ECM) molecules fibronectin, laminin, entactin, and tenascin are glycoproteins that play an important role in the development and function of cells associated with the basement membrane. The distribution of the ECM molecules differs from organ to organ. For example, tenascin has a much more limited distribution than fibronectin, laminin, and entactin and is found more often in developing tissues. All the molecules listed above have an RGD sequence that is the cell-binding domain. Fibronectin is the only one of the four listed ECM molecules found in a circulating plasma form. Not all ECM molecules have an identical structural pattern. Laminin, entactin, and fibronectin have collagen-binding domains, while tenascin does not appear to bind to collagen. ECM molecules interact with cellular receptors called *integrins* (e.g., fibronectin or laminin receptors).

125. The answer is A. *(Junqueira, 8/e, p 98. Uitto, pp 9, 109, 111, 114, 126.)* Ehlers-Danlos disorders include type IV disease in which there are problems in the transcription of type III collagen mRNA or in translation of this message. The result is breakdown of the type III collagen, which is responsible for the elasticity of the intestinal and aortic walls. Hyperextensible skin occurs in Ehlers-Danlos type VI disorder in which problems with the hydroxylation of the amino acid lysine and subsequent cross-linking result in enhanced elasticity. Type VII Ehlers-Danlos disorder involves a specific deficiency in an amino terminal procollagen peptidase. This results from a genetic mutation that alters the propeptide sequence in such a way that the molecular orientation and cross-linking are adversely affected. Increased degradation of proteoglycans occurs in osteoarthritis. Type I collagen is found in dentin.

126. The answer is A. *(Junqueira, 8/e, pp 118–123. Ross, 3/e, pp 126–131. Stevens, pp 54–55.)* Adipose tissue is specialized for lipid storage. It is a low-density method for the storage of large amounts of energy. Adipose tissue also functions as a thermal insulator and a shock absorber. Each fat cell, or adipocyte, is surrounded by a basal lamina. Both brown and white fat cells are most likely derived from a similar precursor, an undifferentiated mesenchymal cell.

White adipose tissue is unilocular and the cells have a single, large lipid droplet in the cytoplasm that provides the "signet-ring" appearance often described for fat cells. White fat cells contain numerous pinocytotic vesicles. Brown adipose tissue has a multilocular appearance and a brown color because of the many mitochondria in these adipocytes. Both types of fat tissue are highly vascularized and function in protection from the cold. Brown fat specifically is involved in heat production while white fat is a true thermal insulator. The former is found in hibernating animals and neonatal humans.

Metabolism of lipid is regulated by neurotransmitters and hormones such as norepinephrine and insulin. This is true of both forms of adipose tissue. In white adipose tissue, there is innervation of the blood vessels by sympathetic fibers rather than direct innervation of the adipocytes. Norepinephrine activates the cyclic AMP cascade through adenylate cyclase. The cyclic AMP activates hormone-sensitive lipase, which removes triglycerides from the stored lipid and hydrolyzes free fatty acids. In white adipocytes, the released fatty acids and glycerol are exported from the cells and combined with albumin, which serves as a blood transport molecule. In brown adipose tissue the fatty acids are used within the cell. However, the electron transport system is uncoupled from oxidative phosphorylation, which results in the production of heat instead of ATP. Heat is transferred to the blood by the extensive capillary networks found in brown adipose tissue.

Shivering initiates the mobilization of lipid in white adipose tissue because shivering requires energy. Lipid mobilization in brown adipose tissue is described as "nonshivering thermogenesis."

127. The answer is E. *(Junqueira, 8/e, pp 99–100. Ross, 3/e, p 102.)* The reticular fibers are very thin compared with collagen or elastin and are visible only in special preparations such as silver stains; therefore, they are known as *argyrophilic* (*silver-loving*). They are also stained with periodic acid–Schiff (PAS) stain because of the high concentration of glycoproteins compared with collagen. Reticular fibers appear in organs during embryogenesis and during repair processes, such as after wounding. They are composed of type III collagen, which interacts with the glycoproteins and proteoglycans, and therefore they have a specific antigenicity and structure compared with type I collagen. The distribution of reticular fibers is specific to hematopoietic (e.g., bone marrow), lymphoid (e.g., lymph node) and glandular (e.g., liver and pituitary) organs as well as smooth muscle. Both reticular and type I collagen fibers are composed of collagen fibrils and exhibit cross-banding.

128.   The answer is E. (*Junqueira, 8/e, pp 88–89. Ross, 3/e, p 94. Stevens, p 42.*) Connective tissue—one of the basic tissues of the body—is so named because it connects epithelium to underlying tissues. It forms the capsule of organs that separates them from the external environment, and it also provides support and mediates cell migration. The connective tissue is composed of fibers, ground substance, and cells. The cells include mast cells, neutrophils, eosinophils, plasma cells, lymphocytes, and macrophages that function in body defense mechanisms. The extracellular matrix is produced by connective tissue cells but also by epithelial and smooth muscle cells.

129.   The answer is D. (*Alberts, 3/e, pp 995–1000. Widnell, pp 268–269.*) The fibronectin receptor is the most well-studied member of the class of transmembrane linkers known as *integrins*. These glycoproteins consist of heterodimers of α- and β-chains. These molecules have a common β-chain but differ in their α-chain. The integrins recognize RGD cell-binding sequences. Through interactions with talin, the fibronectin receptor links the cytoskeleton with molecules in the extracellular matrix such as fibronectin. Receptors for fibronectin are therefore instrumental in regulation of cell motility. Integrins on the surface of blood platelets bind fibrinogen and fibronectin and therefore play an important role in blood clotting. The deficiency of these extracellular matrix receptors occurs in Glanzmann's disease and results in excessive bleeding because of the delay in normal clotting.

130.   The answer is A. (*Alberts, 3/e, pp 973–978. Cotran, 5/e, pp 250–252. Junqueira, 8/e, pp 88–91, 148. Stevens, pp 42–43. Widnell, pp 146–147.*) Glycosaminoglycans are large molecules composed of alternating disaccharide units. The two classifications are hexuronic and hexosamine. Glucosamine and galactosamine are members of the hexosamine category, while glucoronic and iduronic acid are the hexuronic acids. *N*-Acetylglucosamine and *N*-acetylgalactosamine are often sulfated, while uronic acid contains a carboxyl group. The presence of sulfate and carboxyl groups creates a negative charge, which attracts cations such as sodium and leads to hydration and the maintenance of a large space in the extracellular compartment. This water retention leads to resistance to compression by external forces. The large volume of water will facilitate the filtration of materials through the extracellular matrix. Proteoglycans and hyaluronic acid weave together to form pores, a configuration that impedes the flow of larger molecules. The glycosaminoglycans are hyaluronic acid, chondroitin sulfate, dermatan sulfate, heparan sulfate, heparin, and keratan sulfate. Covalent binding to a core protein to form proteoglycan subunits is characteristic of all the glycosaminoglycans except hyaluronic acid, which forms the skeleton for the formation of the proteoglycan aggregate as individual proteoglycan subunits insert into hyaluronic acid. Proteoglycans exhibit a high degree of viscosity and are excellent lubricants because of their ability to undergo extensive hydration. They are also associated with growth factors, such as fibroblast growth factor (FGF), and other extracellular matrix components, such as fibronectin and laminin.

Cellular migration is a complex process involving collagen, laminin, fibronectin, glycosaminoglycans, proteoglycans, and growth factors. Synthesis of hyaluronic acid followed by hydration of the extracellular matrix is a prerequisite for establishment of an appropriate environment for cell migration. Therefore, hyaluronic acid and other glycosaminoglycans are important in the regulation of cell migration.

Other proteoglycans, such as syndecan, are membrane-bound in comparison to the extracellular proteoglycans and play a role in cellular aggregation and the formation of epithelial sheets.

131. The answer is C. (*Alberts, 3/e, pp 1269–1270. Cotran, 5/e, pp 250–252. Isselbacher, 13/e, p 1817.*) Tumor metastasis is like many steps in embryonic development in that it involves cell-cell recognition, migration, and differentiation. Tumor cells are initially released from adhesion to each other and to the extracellular matrix. Dissolution of basement membrane is required for release of tumor cells from the source and passage through connective tissues and between endothelial cells of the blood or lymphatic vessels. Collagenases and other extracellular proteases are involved in this process. Antibodies to integrins (e.g., laminin receptors) and specific collagenases have been shown to block metastatic processes in experimental systems. At the site of a new metastasis, there is a reestablishment of cell-cell and cell-matrix interactions. Very few of the tumor cells released into the bloodstream have metastatic capability or are successful in the production of a tumor at a new site. This has been proved in subcloning experiments that delineated the variation in metastatic potential of cells derived from a specific tumor.

132. The answer is E. *(Alberts, 3/e, pp 980–981. Junqueira, 8/e, pp 98, 116.)* Scurvy, or vitamin C deficiency, results in slower secretion of collagen from fibroblasts. The collagen formed is not normally hydroxylated at proline and lysine residues because of the absence of vitamin C, which is a specific cofactor for hydroxylation of proline and lysine. The result is an inability to form normal triple helices. In scurvy, the resulting collagen is less stable and is subject to denaturation and proteolytic breakdown. Bone growth, development of the dentition, and wound and fracture healing as well as the general stability of adult organs may be affected because of the importance of collagen in the maintenance of structural stability. Periodontal bleeding and ulceration are also common symptoms.

133. The answer is E. *(Alberts, 3/e, pp 115–116, 984–985. Goodman, pp, 193, 197. Junqueira, 8/e, pp 100–101. Ross, 3/e, pp 102–104. Stevens, pp 45–46.)* Elastic fibers allow stretch and recoil in blood vessels (e.g., the aorta and large arteries). The primary component of elastic fibers is elastin, a protein similar to collagen, but containing lower levels of lysine and proline, and showing an absence of glycosylation. The elastin molecules consist of alternating hydrophobic and cross-linking segments. The cross-linked segments are similar to those in collagen, while the hydrophobic segments are responsible for the elasticity of elastic fibers. In the synthesis of elastic fibers, the first stage is the synthesis of a fibril-forming glycoprotein called *fibrillin.* The microfibril arrangement of fibrillin forms a scaffolding for the assembly of elastin molecules. Elastin is secreted as tropoelastin, which is similar to the formation of and secretion of tropocollagen during collagen synthesis. The fibrillin framework, secreted first, allows the tropoelastin monomers to aggregate in random coil form. As humans age, the fibrillin deteriorates, leaving more elastin. Elastic and collagen fibers are synthesized by fibroblasts, but elastic fibers are also synthesized by smooth muscle cells such as those located in the media of large arteries and the aorta.

134. The answer is D. *(Alberts, 3/e, pp 993–995.)* The extracellular matrix is degraded by two groups of proteases: metalloproteases and serine proteases. Collagenase is one of the metalloproteases. Urokinase-type plasminogen activator (U-PA) converts the inactive molecule plasminogen to the active protease plasmin. Tissue inhibitors of metalloproteases (TIMPs) and serpins are responsible for the inactivation of metalloproteases and serine proteases.

135. The answer is A. *(Alberts, 3/e, pp 978–985.)* The fibrillar collagens establish tensile strength at a number of levels including intra- and intermolecular cross-links. Covalent bonding occurs through the OH⁻ groups of hydroxylysine and hydroxyproline and serves to stabilize the triple helix. The triple helix itself functions to resist tensile forces. The degree of cross-linking varies from tissue to tissue. For example, it is highly extensive in tendons. The organization of collagen in tissues also varies depending on function, from the layered appearance in bone to the axial parallel bundles in tendons and the wickered pattern in skin. The interactions with fibril-associated collagens is also important in establishing tissue organization and flexibility.

136. The answer is B. *(Alberts, 3/e, pp 990–992. Goodman, pp 198–200. Junqueira, 8/e, pp 61, 63, 91. Stevens, p 47.)* Laminin is a glycoprotein and a major component of all basement membranes. Both laminin and fibronectin are found in the lamina rara of the basal lamina, but probably on opposite sides of the lamina densa since laminin appears to bind to epithelial cells and fibronectin to the underlying connective tissue cells. One molecule of laminin binds to one molecule of entactin. The binding of entactin to cells through its RGD sequence tightens the binding of cells to the basement membrane. Laminin contains both RGD and YIGSR cell-binding sites as well as collagen and heparan sulfate proteoglycan. The highly charged glycosaminoglycans are responsible for the filtration characteristics of the basement membrane (e.g., renal glomerular basement membrane).

137. The answer is E. *(Alberts, 3/e, pp 976–978.)* The proteoglycans subserve a number of functions in the extracellular matrix. They provide hydration, support, filtering action, cell adhesion, and binding of growth factors. The binding of these proteins may cause immobilization, protection from proteolytic agents, concentration, and sterical blockade of the protein. Proteoglycans may protect growth factors from degradation or inhibitors of degradation (such as plasminogen activator and plasminogen activator inhibitor), but they do not function in the degradation directly.

138. The answer is D. *(Junqueira, 8/e, pp 103–106, 222–223, 226–228. Roitt, 3/e, p 2.13. Ross, 3/e, p 194.)* Macrophages are motile cells, derived from blood monocytes, that secrete and respond to cytokines. For example, macrophages are known to respond to a specific chemoattractant (i.e., MCP-1, chemoattractant protein 1) and colony-stimulating (i.e., colony-stimulating factor) cytokines and secrete other cytokines and growth factors (i.e., the interleukins). Macrophages are known phagocytic cells. Included in this family are Kupffer cells of the liver, microglia of the brain, and Langerhans cells of the skin. The first cells to enter an acute inflammatory site are the neutrophils, also known as *polymorphonuclear leukocytes* because of their nuclear shape. These cells specifically phagocytose bacteria.

139. The answer is B. *(Isselbacher, 13/e, pp 1630–1632. Male, 2/e, pp 12, 94. Roitt, 3/e, pp 2.18–2.19. Ross, 3/e, pp 110–111.)* Mast cells are bone marrow–derived cells that contain heparin, histamine, slow-reacting substance of anaphylaxis (SRS-A), leukotrienes, and eosinophil chemotactic factor of anaphylaxis (ECF-A). Heparin functions as an anticoagulant. Histamine and SRS-A increase vascular permeability. ECF-A is a chemoattractant for eosinophils that contain histaminase, an enzyme that counteracts the effectiveness of secreted histamine. The leukotrienes induce slow smooth muscle contraction. They do not contain or secrete IgE. However, IgE binds to high-affinity receptors on mast cells after the first exposure to an antigen. During the second exposure, the antigen binds to the IgE on the mast cell, resulting in exocytosis of mast cell granules.

140. The answer is C. *(Isselbacher, 13/e, pp 1636–1637. Junqueira, 8/e, pp 106–107.)* Mastocytosis is a disease in which there is an excessive production of mast cells by the bone marrow. The result is an excessive release of the products contained in mast cell granules: histamine, heparin, eosinophil chemotactic factor of anaphylaxis (ECF-A), slow-reacting substance of anaphylaxis (SRS-A), and leukotrienes. Excessive production of acid by the parietal cells of the stomach occurs because of the overstimulation of histamine receptors on these cells. This can result in peptic ulcers and gastritis. Periportal fibrosis of the liver often occurs in systemic mastocytosis due to the extensive infiltration of mast cells. Urticaria pigmentosa which consists of edema (caused by the increased vascular permeability induced by histamine and SRS-A) and infiltration of eosinophils (attracted by ECF-A) which causes itching are also characteristics of mastocytosis..

141–143. The answers are 141-H, 142-E, 143-B. *(Alberts, 3/e, pp 893–894. Cotran, 5/e, pp 85–89.)* Wound healing is a complex process initiated by damage to capillaries in the dermis. The epidermis is avascular and nutrients reach it by diffusion from the dermal capillaries. The clot forms through the interaction of integrins on the surface of blood platelets with fibrinogen and fibronectin. Fibrin is the primary protein that constructs the three-dimensional structure of the clot. Macrophages and fibroblasts are attracted by platelet-derived growth factor (PDGF) and probably by fibronectin. PDGF also stimulates smooth muscle cells, which are involved in the repair of blood vessels, and plays a major role in proliferation of fibroblasts and the simulation of extracellular matrix protein synthesis by fibroblasts at the wound site. A scar is formed as a very dense region of connective tissue fibers. Macrophages remove debris at the wound site and are also involved in the remodeling of the scar. Other growth factors are involved in the wound healing process. Fibroblast growth factor (FGF) and angiogenic growth factor stimulate proliferation and protein synthesis by endothelial cells and fibroblasts. Other steps in wound healing are probably regulated by other growth factors and cytokines.

144–148. The answers are 144-C, 145-A, 146-B, 147-F, 148-D. *(Alberts, 3/e, p 983. Junqueira, 8/e, pp 93–100, 126. Mayne, pp 94–97, 195–221. Stevens, pp 43–45.)* Collagen is a family of extracellular matrix proteins, all of which contain three α-chains that vary in structure and gene expression. There are over twelve defined types. Type I collagen, the most abundant type, is found in bone and fibrocartilage and serves in resistance to tensile stresses. It demonstrates a distinct 67 nm (670 Å) periodicity.

Type II collagen forms fibrils, but not as thick as those formed in type I collagen. It is found in hyaline and elastic cartilage. Electrostatic interactions between type II collagen and proteoglycan aggregates form the molecular basis for the rigidity of hyaline cartilage.

Type III collagen stains in silver preparations and forms the collagenous component of reticular fibers. The reticular fibers form the support for lymphoid organs such as the spleen, bone marrow, and lymph nodes.

Type IV collagen is a major component of the basement membrane. It functions in filtration and is selective because of its interaction with the heparan sulfate proteoglycan to produce a highly polyanionic charge distribution. Type IV collagen is synthesized by endothelial and epithelial cells and retains the propeptide portions of the molecules, which provide the basis for interaction and formation of a sheetlike meshwork. It also interacts with glycoproteins (e.g., fibronectin and tenascin).

Types IX, XII, and XIV collagen are the fibril-associated collagens with interrupted triple helices (FACIT). Collagen fibrils are connected to other extracellular matrix molecules through the FACIT collagens. Type IX collagen binds to type II fibrils while type XII molecules bind to type I collagen. These FACIT collagens regulate the orientation and therefore the function of the fibrillar collagens. The FACIT collagens also retain their propeptides.

Type V collagen is found in placental basement membrane as well as in smooth and skeletal muscle. It may serve a linkage function in the basement membrane.

149–153. The answers are 149-H, 150-B, 151-F, 152-G, 153-E. *(Junqueira, 8/e, pp 101–111. Roitt, 3/e, pp 1.3, 2.1–2.10, 2.17–2.18, 19.6– 19.7. Ross, 3/e, pp 106–111. Stevens, pp 50–51, 71, 74–77.)* The cells of connective tissue include resident cells and migrating cells. The developmental source of these cells also varies with some cells derived from undifferentiated mesenchymal cells and others from the bone marrow. These cell types include fibroblasts, macrophages, mast cells, adipocytes, neutrophils, eosinophils, basophils, plasma cells, and lymphocytes. The fibroblast is the most common connective tissue cell. Fibroblasts are responsible for the synthesis of the fiber (collagen, elastic, and reticular) and ground substance (glycoproteins and proteoglycans) constituents of the connective tissue matrix.

Macrophages are phagocytic cells that originate in bone marrow, pass through the bloodstream as monocytes, and ultimately enter tissue. They phagocytose bacteria and viruses, a process initiated by complement and IgG. Macrophages phagocytose antigen and secrete it onto their cell surfaces, where it is presented to other cells, including T and B lymphocytes. In different tissues the phagocytes or macrophages may have different names that reflect an independent discovery by a scientist and therefore an eponym for the cell type (e.g., Kupffer cells in the liver, Langerhans cells in the skin, and Hofbauer cells in the placenta). Macrophages also respond to and secrete intercellular regulatory molecules called *cytokines.*

Mast cells stain with toluidine blue to form a metachromatic staining product. They are subclassified into two types: (1) the connective tissue mast cell (CTMC), which secretes heparin, is found in the connective tissue, and functions independently of T cells; and (2) the mucosal mast cell (MMC), which is located in the mucosa (close to the lumen or surface), secretes chondroitin sulfate rather than heparin, and is dependent on T lymphocytes. Connective tissue mast cells have IgE on their surfaces and secrete histamine, heparin, leukotrienes, and chemoattractants called *eosinophil-chemoattractant factors (ECFs).*

Eosinophils are phagocytes that appear to be specific for antigen-antibody complexes. They also secrete histaminase, which degrades histamine released from basophils and mast cells, and arylsulfatases, which break down leukotrienes. By carrying out these functions, eosinophils modulate the mast cells and basophils that respond during allergic reactions and provide a means of negative feedback regulation of allergic responses. They are attracted to a site of inflammation by eosinophil-chemoattractant factors (ECFs) released by mast cells and basophils.

Neutrophils (polymorphonuclear leukocytes) function in the killing of bacteria and aggregate in the area invaded by bacteria. As the cells die, they form one of the major constituents of pus.

Basophils are derived from a different stem cell than the mast cells but closely resemble mast cells in structure. They are the only blood source of histamine.

Plasma cells are derived from B lymphocytes after initiation of the humoral response by macrophages, which phagocytose and present antigen (antigen-presenting cells) and T-helper cells, which are required for plasma cell differentiation. Plasma cells are not blood-borne cells, but they secrete antibodies (immunoglobulins) into the bloodstream. They produce all the immunoglobulins: IgG, IgA, IgM, IgD, and IgE.

Lymphocytes are observed in connective tissue. These cells may recirculate through the body and home to specific regions based on homing receptors on their cell surface. These receptors mature on T cells and B cells during the education of these cells in the thymus and bone marrow, respectively. (The circulation and homing of lymphocytes will be covered in greater detail in the chapter on lymphoid tissue.)

Specialized Connective Tissues: Bone and Cartilage

DIRECTIONS: Each question below contains five suggested responses. Select the **one best** response to each question.

154. Hyaline cartilage, which is the most common form of cartilage,

(A) is a highly vascular tissue
(B) is surrounded by a perichondrium except at articular surfaces
(C) contains a matrix composed primarily of glycoproteins
(D) contains type I collagen
(E) is capable of extensive turnover and repair

155. Intramembranous and endochondral ossification differ in

(A) the action of osteoblasts
(B) the light microscopic appearance of the adult bone
(C) the ultrastructural appearance of the adult bone
(D) the presence of woven bone only in intramembranous ossification
(E) the microenvironment in which ossification occurs

156. After birth, growth in the length of long bones occurs through

(A) the periosteal collar
(B) the primary ossification center
(C) the secondary ossification center
(D) appositional growth from the periphery
(E) interstitial growth in the diaphysis

157. In long bone development, hypertrophied chondrocytes are the source of

(A) parathyroid hormone
(B) calcitonin
(C) alkaline phosphatase
(D) type I collagen
(E) osteopontin

158. The molecular basis for shock absorption and resiliency within articular cartilage is

(A) the electrostatic interaction of proteoglycans with type IV collagen
(B) the ability of glycosaminoglycans to bind anions
(C) the noncovalent binding of glycosaminoglycans to protein cores
(D) the sialic acid residues in the glycoproteins
(E) the hydration of glycosaminoglycans

159. Osteoblasts are correctly characterized by which of the following statements?

(A) They are derived from monocytes
(B) They synthesize high quantities of acid phosphatase
(C) They are involved in matrix formation but not mineralization
(D) They have less rough endoplasmic reticulum than do osteocytes
(E) They possess PTH receptors

160. In the restructuring of the skeleton, bone remodeling

(A) occurs exclusively in adult bone
(B) follows a cycle in which bone formation precedes resorption
(C) results in the formation of cement and frontier lines
(D) occurs independently of soft tissue growth and external forces
(E) only occurs at the epiphyses of long bones

161. Which of the following statements is true of metabolic bone markers used for diagnosis?

(A) Hydroxyproline is a good urinary marker for bone metabolism
(B) Pyridinium cross-links may be used for determination of bone formation
(C) Osteocalcin is a marker of bone resorption
(D) Alkaline phosphatase is the best clinical marker for bone resorption
(E) X-ray is currently the best method for detecting osteoporosis

162. Identify the structure marked by the arrow in the light micrograph of a developing long bone.

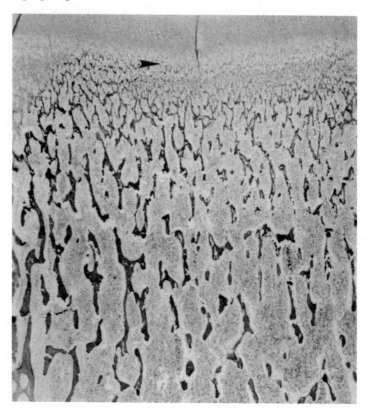

(A) Bone
(B) Calcified cartilage
(C) Hypertrophied chondrocytes
(D) Proliferating chondrocytes
(E) Periosteal bud

163. The cell in the accompanying electron micrograph functions primarily

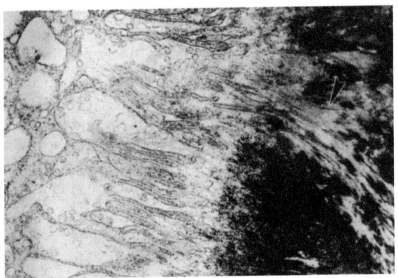

From Erlandsen SL, Magney JE: *Color Atlas of Histology,* CV Mosby, 1992, with permission.

(A) as a source of monocytes
(B) through PTH receptors on its surface
(C) in response to calcitonin
(D) to decrease blood Ca^{2+} levels
(E) in response to low Ca^{2+} levels

164. A patient with rheumatoid arthritis would exhibit which of the following joint changes?

(A) Loss of the proteoglycan matrix and fibrillation in the articular cartilage during the early stages
(B) Decreased levels of fibrinogen in the synovial fluid
(C) Formation of osteophytes at the articular margins and eburnation of large weight-bearing joints in the later stages
(D) Decreased numbers of leukocytes including PMNs in the synovial fluid
(E) Heterologous autoantibodies directed against serum gammaglobulin that are deposited in the synovial membrane and lead to pannus formation

165. In the healing of fractures, a bone graft functions primarily as a

(A) source of osteoblasts, which transform the clot into a callus
(B) source of additional periosteal cells
(C) source of woven bone
(D) source of osteoclasts
(E) temporary bridge in a severe fracture defect

166. A 55-year-old woman presents with pain in her right hip and thigh. The pain started approximately 6 months ago and is a deep ache that worsens when she stands or walks. Your examination reveals increased warmth over the right thigh. The only laboratory abnormalities are alkaline phosphatase 656 IU/L (normal 23 to 110 IU/L), elevated 24-h urine hydroxyproline, and osteocalcin 13 ng/mL (normal <6 ng/mL). X-ray of hips and pelvis shows osteolytic lesions and regions with excessive osteoblastic activity. Bone scan shows significant uptake in the right proximal femur.

Which of the following would you include in your differential diagnosis?

(A) Paget's disease
(B) Multiple myeloma
(C) Osteomalacia
(D) Osteoporosis
(E) Hypoparathyroidism

167. A 66-year-old man with no previous significant illness presents with back pain. The patient had felt well except for an increase in fatigue over the past few months. He suddenly felt severe low back pain while raising his garage door. Physical examination reveals a well-developed white male in acute pain. His pulse is 88 beats per minute and blood pressure is 150/90 mmHg. The conjunctivae are pale. There is marked tenderness to percussion over the lumbar spine. The following laboratory data are obtained: hemoglobin 11.0 g/dL (normal 13 to 16 g/dL), serum calcium 12.3 mg/dL (normal 8.5 to 11 mg/dL), abnormal serum protein electrophoresis with a monoclonal IgG spike, urine positive for Bence Jones protein, abnormal plasma cells in bone marrow. X-rays reveal lytic lesions of the skull and pelvis and a compression fracture of lumbar vertebrae.

Your diagnosis would be

(A) osteoporosis
(B) osteomalacia
(C) multiple myeloma
(D) hypoparathyroidism
(E) Paget's disease

168. A 46-year-old woman presents with a pain in the left leg that worsens on weight-bearing. An x-ray shows demineralization and a decalcified (EDTA-treated) biopsy shows reduction in bone quantity. The patient had undergone menopause at age 45 without estrogen replacement. She reports long-standing diarrhea. In addition, laboratory tests show low levels of 25-hydroxyvitamin D, calcium, and phosphorus and elevated alkaline phosphatase. A second bone biopsy, which was not decalcified, shows uncalcified osteoid on all the bone surfaces. On the basis of these data, your diagnosis would be

(A) osteoporosis
(B) osteomalacia
(C) scurvy
(D) Paget's disease
(E) hypoparathyroidism

DIRECTIONS: Each numbered question or incomplete statement below is NEGATIVELY phrased. Select the **one best** lettered response.

169. The stimulation and activation of osteoclast-mediated release of Ca^{2+} from bone involves all the following EXCEPT

(A) mechanical action of osteoclasts grinding the bone matrix
(B) synthesis and secretion of acid phosphatases and collagenase to digest the bone matrix
(C) formation of the equivalent of a secondary lysosome as an acidified resorption compartment
(D) proton pump activity analogous to that of the parietal cells of the stomach
(E) regulation by local cytokines produced by other cell types

170. Patients with Cushing's syndrome often show osteoporotic changes. All the following may be involved in the etiology of osteoporosis induced by Cushing's syndrome EXCEPT

(A) decreased glucocorticoid levels that result in decreased quality of the bone deposited
(B) interference with normal bone deposition
(C) inhibition of intestinal calcium absorption
(D) increased PTH levels
(E) bone fragility resulting from excess bone resorption

171. Osteogenesis imperfecta is a genetic disorder in which the molecular basis is a defect in type I collagen. Which of the following would NOT be an expected sequela of osteogenesis imperfecta?

(A) Weak bones that fracture easily
(B) Osteopenia
(C) Dentinogenesis imperfecta
(D) Reduction in normal levels of woven bone
(E) Blue sclerae

172. DiGeorge syndrome is a congenital malformation in which the embryologic derivatives of the third and fourth branchial pouches fail to form. All the following are likely to occur in children with this syndrome EXCEPT

(A) absence of the parafollicular cells
(B) deficiency of cells in the deep cortex of the lymph nodes
(C) tetany
(D) deficiency in osteoclasts
(E) decreased Ca^{2+} levels in the blood

173. This electron micrograph is a preparation from bone matrix in close proximity to the cellular components. All the following statements are true in regard to these structures EXCEPT

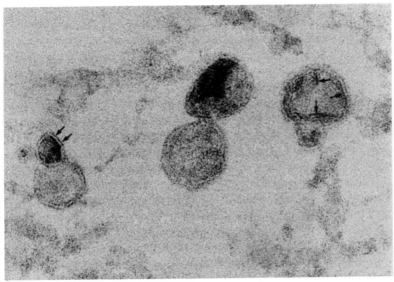

Courtesy of Dr. David Morris.

(A) they accumulate calcium and phosphate
(B) they are absent during endochondral bone formation
(C) they bud off from osteoblasts
(D) they may serve a "seed crystal" function in developing bone
(E) they contain alkaline phosphatase

---

174. The noncollagenous proteins in bone subserve all the following functions EXCEPT

(A) growth factors
(B) binding of ionic calcium and physiologic hydroxyapatite
(C) formation of the three-dimensional fiber structure of the matrix
(D) collagen fibrillogenesis
(E) cell attachment

175. Vitamin D deficiency should be suspected in patients who exhibit any of the following EXCEPT

(A) milk allergy
(B) hypercalcemia and hyperphosphatemia
(C) chronic diarrhea
(D) a history of taking anticonvulsants
(E) a history of taking cholesterol-binding resins

DIRECTIONS: Each group of questions below consists of lettered headings followed by a set of numbered items. For each numbered item select the **one** lettered heading with which it is **most** closely associated. Each lettered heading may be used **once, more than once, or not at all.**

Questions 176–179

Use the diagram of a joint below to match each description with the correct labeled structure.

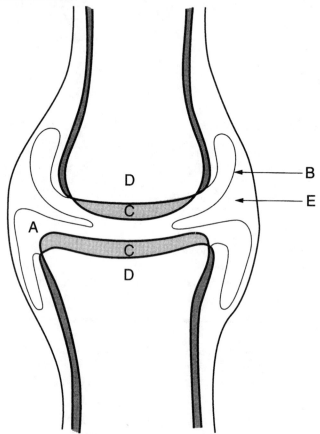

176. The site of macrophage-like cells that phagocytose from the synovial fluid

177. The site of cells that synthesize the synovial fluid

178. The structure replaced by the pannus in rheumatoid arthritis

179. The primary site of injury in osteoarthritis

Specialized Connective Tissues: Bone and Cartilage

Answers

154. The answer is B. *(Burkitt, 3/e, pp 170–172. Erlandsen, p 29. Junqueira, 8/e, pp 124–128. Ross, 3/e, pp 132–135. Stevens, pp 52–53.)* Cartilage, along with bone, forms the specialized connective tissues that serve the supportive function of the skeletal system. Compared with bone, cartilage is an avascular tissue and is dependent upon diffusion to obtain nutrients and remove waste products. Since chondrocytes function through glycolytic metabolic pathways, they have adapted to the anaerobic conditions of the cartilage matrix. Cartilage also exhibits low metabolic activity and repairs poorly after damage in comparison with bone and other tissues. Like other forms of connective tissue, cartilage is comprised of cells, fibers, and ground substance. The cells are chondrocytes, specialized cells that differentiate from osteoprogenitor cells derived from mesenchyme. The ground substance contains predominantly proteoglycans, but also growth factors such as chondronectin, which is important in the adhesiveness of chondrocytes for the matrix. The fibers vary depending upon the type of cartilage.

Cartilage is found in three forms: hyaline cartilage, elastic cartilage, and fibrocartilage (also known as *fibrous cartilage*). Hyaline cartilage is the most common form. In the embryo it is found in all developing long bones, which are formed first as cartilage models and are subsequently replaced by bone through endochondral ossification. In the adult, hyaline cartilage is found in articular cartilages and is important for support in the wall of the respiratory tree, including the trachea through the bronchi. Hyaline cartilage contains type II collagen, elastic cartilage contains type II collagen and elastic fibers, and fibrocartilage contains type I collagen. While hyaline cartilage is surrounded by a perichondrium except at articular surfaces and elastic cartilage is surrounded by a perichondrium, fibrocartilage has no perichondrium. Elastic cartilage is found in the epiglottis and the pinna of the ear. Fibrocartilage is always found in an intermediate or transitional position between hyaline cartilage and dense connective tissue. It is located in the pubic symphyses and intervertebral disks.

155. The answer is E. *(Junqueira, 8/e, pp 139–144. Ross, 3/e, pp 161–166. Stevens, pp 243–244.)* Bone development occurs by two different methods, which differ in location and the presence or absence of cartilage models of the bones. For example, in the flat bones of the skull, bone formation occurs through the differentiation of osteoprogenitor cells from mesoderm and is accompanied by vascularization. This is known as *intramembranous ossification*. In the other form of ossification, osteoprogenitor cells differentiate into chondrocytes and establish a cartilage model of the long bone. This method occurs in bones such as the humerus and femur. The cartilage model of each bone is replaced by bone using the cartilage as a scaffolding for bone formation. In both cases, bone development occurs by essentially the same process, the synthesis of collagen and other matrix components by osteoblasts and the calcification of the matrix through the action of alkaline phosphatase from osteoblasts. The difference between the two mechanisms of bone development is the microenvironment in which they occur. Bone formed by the two methods cannot be distinguished microscopically or macroscopically. In both endochondral and intramembranous ossification, the first bone formed is woven bone (also known as *primary bone*). This bone is replaced by adult, lamellar bone through a remodeling process. Fracture healing repeats many of the steps involved in both forms of bone development.

156. The answer is C. *(Junqueira, 8/e, pp 140–145. Ross, 3/e, pp 164–166. Stevens, p 245.)* Fetal development of long bones occurs by the process of endochondral ossification in which a cartilage model is replaced by bone. Before birth in endochondral ossification, the growth in length of the long bone occurs primarily through the proliferation of chondrocytes in the proliferative zone within the diaphysis of the cartilage model. Growth in the width of the long bone occurs by the addition of osteoblasts from the periosteum. This is essentially a form of appositional growth without a cartilage intermediate. It is one of the best examples of intramembranous ossification, even though it occurs in the development of a long bone. Growth in the length of long bones after birth occurs through cell proliferation of chondrocytes in the epiphysis.

157. The answer is C. *(Junqueira, 8/e, pp 140–145. Stevens, p 244.)* Growth of cartilage occurs by appositional or interstitial growth. Appositional growth occurs by the addition of chondrocytes from the perichondrium. These cells produce matrical components resulting in enlargement of the cartilage. Interstitial growth occurs by the proliferation of chondrocytes from within the matrix. These cells form isogenous groups that represent daughter cells resulting from cell division by immature chondrocytes within the cartilage matrix. Long bone development represents a modification of growth processes as a result of the calcification of the matrix, which prevents appositional growth.

In the development of the long bones, specific zones are established within the cartilage model. Moving from the epiphyseal ends toward the center of the diaphysis, one finds the following zones: resting, proliferation, hypertrophy, calcified cartilage, and calcification. Cells in the proliferative zone are involved in the growth in length of the long bone by the birth of many new cells. These chondrocytes synthesize large quantities of alkaline phosphatase, which results in the calcification of the cartilage and the death of the chondrocytes since the cartilage matrix can no longer support the metabolism of the chondrocytes through diffusion. At this point a periosteal bud grows into the center of the shaft, carrying osteoprogenitor cells and providing vascularity, which is required for bone formation. The differentiating osteoblasts deposit bone on the calcified cartilage, which functions as a framework for bone formation. Bone is formed by the action of osteoblasts forming type I collagen, noncollagenous proteins (e.g., osteopontin and osteonectin), and alkaline phosphatase, which plays an essential role in mineralization of the osteoid. Parathyroid hormone and calcitonin are active during fetal development, but they are secreted by the parathyroid glands and C cells of the thyroid, respectively.

158. The answer is E. *(Alberts, 3/e, pp 975–976. Junqueira, 8/e, pp 124–128. Ross, 3/e, p 133. Stevens, p 52.)* Proteoglycans are the major component of the ground substance of cartilage. They possess a large anionic charge because of the presence of sulfate, hydroxyl, and carboxyl groups within the glycosaminoglycans, which join to form proteoglycan subunits by linking with a core protein. The proteoglycan subunits (monomers) subsequently form an aggregate by linking noncovalently to hyaluronic acid. These aggregates react electrostatically with type II collagen, probably through the sulfate groups of the glycosaminoglycans. The negative charge of the glycosaminoglycans facilitates the binding of cations and the transport of electrolytes and water within the matrix. This is an important aspect of cartilage metabolism since the chondrocytes are dependent on diffusion to obtain nutrients or to dispose of waste products. The hydration of the glycosaminoglycans plays an important role in shock absorption and enhances the resiliency of the cartilage. This role is particularly important in the articular cartilages, which receive pressure during joint movement and are required to resist strong compressive forces. Glycoproteins are not a major constituent of the cartilage matrix.

159. The answer is E. *(Alberts, 3/e, pp 1182–1183. Favus, 2/e, pp 5–6, 15–20. Junqueira, 8/e, pp 132–134. Ross, 3/e, p 157. Stevens, p 236.)* Osteoblasts, the bone-forming cells of the body, are derived from osteoprogenitor cells. They synthesize type I collagen and the noncollagenous bone proteins that make up the uncalcified matrix, or osteoid. In order to carry out this function, the osteoblasts have an extensive rough endoplasmic reticulum. Osteoblasts also synthesize alkaline phosphatase, which is responsible for an increase in the local calcium/phosphate ratio and the subsequent calcification of the matrix. The cells entrapped in the calcified matrix exhibit a decreased metabolic activity, as exemplified by a decreased volume of rough endoplasmic reticulum, and are then called *osteocytes*. The osteoblast is the target cell for parathyroid hormone (PTH) and $1,25(OH)_2$-vitamin D_3, two hormones with anabolic effects on osseous tissues. Acid phosphatase is produced by osteoclasts, not osteoblasts.

160. The answer is C. (*Favus, 2/e, pp 8–9. Junqueira, 8/e, pp 143–145. Stevens, pp 240–241.*) From the time that bone is first deposited in primary ossification centers, it is remodeled through the action of osteoclasts. The skeleton is constantly undergoing restructuring in order to maintain Ca^{2+} levels as well as to maintain normal bone. Remodeling occurs both during development and in the adult with a cycle of activation, resorption, reversal, and bone formation in which bone resorption must precede bone formation. This cycle not only explains turnover of adult bone, but activity in both the epiphyseal plate and in trabecular and Haversian remodeling during the development of long bones. The process results in the presence of cement lines, which represent the initiation of bone formation following bone deposition. The frontier line represents an area of new osteoid production where the mineralization lag has delayed calcification. Bone remodeling also occurs in the flat bones of the skull, which form by intramembranous ossification. In that case reshaping occurs through the deposition of new bone under the periosteum (adjacent to the dura) and resorption along the endosteal surface. The growth of the skull and the development of normal brain size are therefore interrelated. Bone development and remodeling are very sensitive to soft tissue changes and external forces and are regulated by systemic and local factors.

161. The answer is A. (*Coe, pp 1042–1044. Favus, 2/e, p 22.*) There are a number of useful biochemical markers of bone metabolism. Osteoclasts synthesize tartrate-resistant acid phosphatase so that increased osteoclastic activity is reflected in increased serum levels of tartrate-resistant acid phosphatase. Bone resorption fragments of collagen type I and noncollagenous proteins increase as bone matrix is resorbed. Hydroxyproline is a good urinary marker of bone metabolism since hydroxyproline is released and excreted in the urine as collagen is broken down. The presence of pyridinium cross-links (the maturation products of reducible ketoamine cross-links), which are involved in the bundling of type I collagen, is used for measurement of bone resorption. These cross-links are released only during degradation of mineralized collagen fibrils as occurs in bone resorption. Usually pyridinium cross-links are measured by immunoassay over a 24-h period to detect excess bone resorption and collagen breakdown in disorders such as Paget's disease.

Markers of bone formation include osteocalcin, alkaline phosphatase, and the extension peptides of type I collagen. Osteocalcin is a vitamin K–dependent GLA protein that is synthesized by osteoblasts and secreted into the serum in an unchanged state. Serum concentrations of osteocalcin are therefore directly related to osteoblastic activity. It is more specific than the marker alkaline phosphatase, which is less specific because other organs such as the liver and kidney produce this enzyme. The procollagen extensions that assist in fiber formation are released into serum during bone formation. Procollagen III extensions in serum should be measured to correct for nonbone collagen synthesis.

Radiologic methods such as conventional x-ray can be used to detect osteoporosis, but only after patients have lost 30 to 50 percent of their bone mass. Dual-beam photon absorptiometry allows a much more accurate diagnosis of loss of bone mass.

162. The answer is C. *(Burkitt, 3/e, pp 181–185. Erlandsen, pp 36–37. Ross, 3/e, pp 164–165, 176–179.)* The light micrograph illustrates a developing long bone. During development of the long bones of the body, specific zones are established as a cartilage model of a long bone is converted to mature bone. The zones from the epiphysis toward the center of the shaft (diaphysis) are as follows: resting zone, proliferative zone, hypertrophy zone, and zone of calcified cartilage, which is used as the scaffolding for the deposition of bone. The periosteal bud represents the ingrowth of blood vessels, bone marrow, and osteoprogenitor cells into the diaphysis. The zone of hypertrophy contains chondrocytes that have secreted large amounts of alkaline phosphatase. This secretion results in calcification of the matrix and eventual death of these cells, which are dependent on diffusion to obtain oxygen and nutrients from the matrix.

163. The answer is E. *(Alberts, 3/e, pp 1182–1184. Favus, 2/e, pp 68–69. Greenspan, 3/e, pp 247–249. Junqueira, 8/e, pp 133–135. Ross, 3/e, pp 159–160. Stevens, p 238.)* The electron micrograph illustrates the typical ultrastructure of an osteoclast with its distinctive ruffled border. The arrowhead indicates collagen within the degraded bone matrix. Osteoclasts are of hematopoietic origin and appear to be related to the monocytic lineage. They are derived from a monocyte-related cell, but they are not the source of monocytes. Osteoclasts are responsive to a number of hormones including parathyroid hormone (PTH), calcitonin, and $1,25(OH)_2$-vitamin D_3. PTH and dihydroxyvitamin D receptors are found on osteoblasts but not on osteoclasts. The osteoclasts function to raise serum Ca^{2+} by removing calcium from bone. The Ca^{2+} enters the bone fluid, the extracellular fluid, and subsequently the blood. It is currently believed that factors released by osteoblasts in response to PTH increase ruffled-border activity and stimulate the development of osteoclasts. Calcitonin receptors are present on the surface of osteoclasts, but calcitonin is only responsible for transient changes in bone resorption. In the presence of high Ca^{2+}, calcitonin is synthesized and released from C (interfollicular) cells of the thyroid, which decreases ruffled-border activity. Long-term responses to elevated Ca^{2+} are mediated by lower PTH levels rather than increased calcitonin production. This is exemplified in patients with an absence of calcitonin secretion (e.g., after thyroidectomy), or with stimulated levels of calcitonin (e.g., in medullary carcinoma of the thyroid) who exhibit relatively normal bone metabolism. However, calcitonin measurements are an important tool in the diagnosis of medullary thyroid carcinoma.

164. The answer is E. *(Cotran, 5/e, pp 1247–1248. Stevens, p 248.)* Arthritis involves inflammatory changes in a joint. Osteoarthritis begins with loss of hydrated glycosaminoglycans, followed by death of chondrocytes, fibrillation, and development of fissures in the cartilage matrix. The severe wear and tear of osteoarthritis increases with age. During the breakdown of the articular cartilages, the width of the underlying bone increases. Osteoarthritis typically includes the formation of reactive bone spurs called *osteophytes,* which may break off to form foreign bodies in the joint space (i.e., "joint mice"). In the fingers, osteoarthritis primarily affects distal interphalangeal joints, where it produces painful nodular enlargements called *Heberden's nodes.* Large weight-bearing joints are also usually involved in osteoarthritis and often exhibit eburnation in the late stages when the articular cartilages have been worn down and result in an osseous articular surface.

Rheumatoid arthritis is an autoimmune disease in which a rheumatoid factor composed of heterologous autoantibodies directed against serum gammaglobulin (IgG) appears. Rheumatoid factor is present in the serum of 85 to 90 percent of patients with rheumatoid arthritis. It has been suggested that the deposition of rheumatoid factor, which contains complement, can be pathogenic and lead to inflammatory destruction of the joint surface; rheumatoid arthritis thus can represent a sort of immune complex disease. Cell-mediated immunity is also involved in rheumatoid arthritis. Alteration of the synovial membrane results in the formation of a pannus, or inflammatory, hypertrophic synovial villus. The presence of the pannus and release of lysosomal enzymes from the pannus result in degradation of the cartilage. This is followed by hypertrophy and hyperplasia of the articular cartilages, which often leads to bone formation across the joint with welding of the bones together (ankylosis). Due to the inflammation in rheumatoid arthritis, there are elevated numbers of leukocytes, in particular PMNs in the synovial fluid. Fibrinogen, another indicator in synovial fluid of inflammatory responses, increases during rheumatoid arthritis.

165. The answer is E. *(Junqueira, 8/e, pp 145–146.)* In the healing of fractures, the first step is clotting of extravasated blood. The clot is organized into a callus by granulation tissue that consists of fibroblasts, osteogenic cells, and budding capillaries. An internal (bony) callus forms where local bone factors are most active (i.e., in close proximity to the periosteum and endosteum, which retain osteogenic potential). An external (cartilaginous) callus forms bone by endochondral ossification following initial chondrogenesis. These steps involve repetition of the cellular events involved in the histogenesis of bone. A bone graft is not a major source of cells or bone but a method of forming a temporary bridge in a severe defect. Other methods useful in stimulating bone repair include electrical forces and bone morphogenetic protein, a bone growth factor obtained from decalcified bone matrix. This protein stimulates bone formation when implanted at the fracture site.

166. The answer is A. *(Coe, pp 1042–1052. Cotran, 5/e, pp 1223–1225. Greenspan, 3/e, pp 311–314.)* Paget's disease is also known as *osteitis deformans* because of its deforming capabilities (e.g., skull or femoral head enlargement). In this disease the serum calcium is normal, but there is an increase in osteoclastic activity (osteolytic lesions and elevated 24-h urine hydroxyproline) and an increase in osteoblastic activity (elevated osteocalcin and alkaline phosphatase). Patients with Paget's disease exhibit a marked increase in osteoid, and the bone actually enlarges. The osteoid is never normally mineralized in this disease. In this patient the bone scan shows significant uptake of labeled bisphosphonates, which are incorporated into newly formed osteoid during bone formation. Her proximal femur is enlarged and no longer fits properly into the acetabulum, which results in the hip pain.

167. The answer is C. *(Alberts, 3/e, p 1216. Coe, pp 803–804. Cotran, 5/e, pp 663–664.)* The patient is suffering from multiple myeloma. In this disease, there are abnormal changes in the bone marrow indicative of altered plasma cell activity and anemia (hemoglobin data and increasing fatigue). These plasma cells produce elevated levels of interleukin 1 (IL-1), which functions as an osteoclast activation factor. The increased IL-1 stimulates osteoclastic activity and results in elevated serum calcium (12.3 mg/dL). The depletion of bone calcium results in lytic lesions of the skull and pelvis as well as the presence of the compression fracture of the spine. The Bence Jones protein represents free immunoglobulin light chains, which occur in the urine of patients with multiple myeloma.

168. The answer is B. *(Cotran, 5/e, pp 414–416, 1225–1226. Greenspan, 3/e, pp 298–303. Junqueira, 8/e, pp 146–147.)* The patient suffers from osteomalacia, a disease related to malnutrition, specifically vitamin D deficiency. On the basis of the first bone biopsy in which the tissue was decalcified, one could make a diagnosis of osteoporosis. The second, nondecalcified bone biopsy indicates that osteoid is being formed but is not undergoing mineralization. This fits with the low 25-hydroxyvitamin D levels. Vitamin D replacement and calcium supplementation would be prescribed for this patient.

169. The answer is A. *(Alberts, 3/e, pp 1182–1184. Favus, 2/e, pp 6–8. Junqueira, 8/e, pp 132–135. Stevens, p 238.)* Osteoclasts are responsible for the resorption of bone. Although the osteoclast contains some vesicles and vacuoles that enclose bony matrix particles, this cell functions through the production of an acidic environment for the breakdown of bone. Osteoclasts are giant, multinucleate, motile cells with a foamy cytoplasm. The numerous lysosomes, vesicles, and vacuoles contain acid hydrolases that produce an acidic environment for the breakdown of the bony matrix. These enzymes are targeted for secretion by the presence of mannose-6-phosphate receptor. The plasmalemma of the osteoclast adjacent to the resorbing bone surface is thrown into folds and villouslike processes whose tips reach and even enter the bone surface. This surface is called the *ruffled border.* The osteoclast is attached to the bone surface and the resorption area is sealed off by the presence of contractile proteins in the cytoplasm lateral to the site of the ruffled border. The basolateral membrane of the osteoclast possesses an Na^+,K^+-ATPase pump. The osteoclast acidifies the microenvironment beneath the ruffled border by pumping protons provided in the cytosol from carbonic anhydrase. The bone compartment around the ruffled border of the osteoclast is therefore analogous to a secondary lysosome since it contains (1) an acidic environment to facilitate the dissolution of hydroxyapatite crystals, (2) lysosomal enzymes, and (3) an enzymatic substrate (i.e., the bone matrix).

170. The answer is A. *(Coe, pp 831–856. Greenspan, 3/e, pp 305–311.)* Osteoporosis is a major problem of normal aging in both sexes but is particularly prevalent in older women. In this disease, the quality of bone is unchanged, but the balance between bone deposition and bone resorption is lost. The disease is prevalent in postmenopausal women because the protective effect of estrogens is no longer present. Many of the symptoms of osteoporosis may be alleviated by hormonal replacement therapy with estrogen. Osteoporosis may also be induced by other diseases (e.g., hyperthyroidism) or drugs (e.g., alcohol and caffeine). In addition, excess glucocorticoids induce osteoporosis. For example, in Cushing's syndrome patients produce high levels of corticosteroids, which results in interference with bone metabolism. A similar pattern may be seen during prolonged steroid therapy. The result is increased bone resorption compared with bone deposition. Intestinal calcium absorption is inhibited and PTH levels may be increased.

171. The answer is D. *(Alberts, 3/e, p 982. Favus, 2/e, pp 11–12, 238–239. Isselbacher, 13/e, pp 2111–2113.)* Osteogenesis imperfecta is an autosomal dominant or recessive disease with several distinct subtypes with symptoms varying from mild to severe. Persons with this disease have osteopenia with weak bones that fracture easily, hence the synonym, *brittle bone disease.* The mutation is in the collagen I gene and often affects other organs such as the sclera, which appears blue because of the thinness of the collagen layers. The teeth may also be affected with production of defective dentin, which also contains type I collagen. Due to the constant fracture healing there is an increase in woven bone, the first bone formed by osteoblasts. In fact, there is an increase in the number of osteoblasts compared with normal bone.

172. The answer is A. *(Junqueira, 8/e, pp 251, 256, 258, 265–266, 404– 405. Larsen, p 162. Moore, Developing Human, 5/e, p 199. Ross, 3/e, pp 343, 346, 348–349, 606–609.)* DiGeorge syndrome is a congenital malformation that results in the absence of the thymus and parathyroid glands, which arise from the third and fourth pairs of branchial pouches. The absence of the thymus results in a deficiency in T-lymphocyte–dependent areas of the immune system. These areas include the deep cortex of the lymph nodes, periarterial lymphatic sheath (PALS) of the spleen, and interfollicular areas of the Peyer's patches. Parathyroid hormone (PTH) stimulates the development of osteoclasts and the formation of ruffled borders in osteoclasts. The absence of PTH results in (1) a drastic reduction in osteoclasts and activation of osteoclasts as exemplified by the differentiation of ruffled borders, (2) reduced Ca^{2+} levels in the blood, (3) denser bone, (4) spastic contractions of muscle called *tetany,* and (5) excessive excitability of the nervous system. The parafollicular (C) cells arise from the ultimobranchial body that migrates into the developing thyroid gland.

173. The answer is B. *(Junqueira, 8/e, p 143. Ross, 3/e, pp 157, 168. Stevens, pp 238–239.)* The electron micrograph includes matrix vesicles that are derivatives of osteoblast, hypertrophied chondrocyte, ameloblast, and odontoblast cell membranes. After budding off from the plasmalemma, matrix vesicles accumulate calcium and phosphate in two steps. The first step involves intracellular accumulation under the influence of acidic phospholipid Ca^{2+}-binding proteins and phosphatases. This leads to the development of intracellular hydroxyapatite crystals (indicated by the arrows), which in the second step become exposed to the extracellular fluid and lead to seeding of the osteoid between the spaces in the collagen fibrils located in the matrix. Matrix vesicular alkaline phosphatase results in local increases in the Ca^{2+}/PO_4^{2-} ratio. Adult lamellar bone contains very few matrix vesicles. The three-dimensional arrangement of collagen with the presence of holes or pores where hydroxyapatite crystals form is involved in the mineralization of adult bone. The presence of the noncollagenous bone proteins, including the specific binding proteins osteocalcin and osteonectin, has also been implicated in the regulation of mineralization of adult bone.

174. The answer is C. *(Favus, 2/e, pp 22–23. Junqueira, 8/e, pp 135, 143.)* The noncollagenous bone proteins are primarily synthesized by osteoblasts and constitute 10 to 15 percent of bone protein. Some plasma proteins are preferentially absorbed by the bone matrix. In comparison, type I collagen, which is also synthesized by osteoblasts, accounts for 85 to 90 percent of total bone protein. The collagen is responsible for the three-dimensional fiber structure of the matrix. The noncollagenous proteins include cytokines and growth factors, which are synthesized endogenously and become trapped in the matrix. Also included in the category of noncollagenous proteins are the cell attachment proteins (fibronectin and osteopontin); proteoglycans (e.g., chondroitin 4-sulfate and chondroitin 6-sulfate), which appear to play a role in collagen fibrillogenesis; and the *gla* proteins, such as osteocalcin (containing gamma-carboxyglutamic acid), which binds Ca^{2+} and mineral components to the matrix.

175. The answer is B. *(Isselbacher, 13/e, pp 2144–2145.)* Vitamin D deficiency may result from a number of different causes: (1) insufficient dietary supplementation because of milk allergy; (2) drugs like anticonvulsants and cholesterol-binding resins that interfere with vitamin D production or metabolism; (3) chronic diarrhea, resulting in inadequate absorption from the small intestine; and (4) lack of exposure to sunlight. Ultraviolet B radiation is one means of conversion of 7-dehydrocholesterol to previtamin D_3. Lack of vitamin D leads to insufficient absorption of calcium from the small intestine and a decrease in phosphate absorption as well. Low calcium levels lead to an increase in parathyroid hormone (PTH) levels in an attempt to compensate for the hypocalcemia. The elevated PTH levels balance the hypocalcemia to a small degree, but also decrease renal tubular reabsorption of phosphate, leading to hypophosphatemia.

176–179. The answers are 176-B, 177-B, 178-C, 179-C. *(Burkitt, 3/e, pp 187–189. Junqueira, 8/e, pp 147–151. Stevens, pp 246–248.)* Joints are classified as those that are freely movable (diarthroses) and those with limited or no movement (synarthroses). Synarthroses are subclassified as united by bone (synostoses), united by hyaline cartilage (synchondroses), and united by dense connective tissue of a ligament (syndesmoses).

The joint shown in the figure is a diarthrosis. The joint capsule consists of an epithelium (B) and an external fibrous layer (E). The synovial fluid is formed from the synovial capillary ultrafiltrate as well as mucins, hyaluronic acid, and glycoproteins produced by fibroblast-like cells in the synovial epithelium (B) that lines the fluid-filled synovial cavity (A). Macrophage-like cells in the epithelium carry out a phagocytic function. Synovial fluid, which differs from blood serum in its reduced protein content, acts as a lubricant and becomes more viscous with age. It may be used to diagnose joint disorders such as arthritis.

The ends of the bone (D) are covered by hyaline cartilage that lacks a perichondrium. These cartilaginous structures are called the *articular cartilages* (C) and are the primary site of destruction in osteoarthritis. Rheumatoid arthritis is an autoimmune disease in which infiltration of cells from the immune system leads to the destruction of the synovial capsule and the articular cartilages. The pannus is a fibrocollagenous structure that replaces the articular cartilage during the onset of rheumatoid arthritis.

Muscle and Cell Motility

DIRECTIONS: Each question below contains five suggested responses. Select the **one best** response to each question.

180. Which of the following statements is true of the organization of skeletal muscle and its relationship to connective tissue?

(A) Individual muscle fibers are surrounded by basal lamina-like connective tissue called *perimysium*

(B) Specialized connective tissue functions in the mechanical transmission of the force of contraction

(C) Groups of muscle fibers are held together by connective tissue called *epimysium*

(D) The connective tissue found immediately between muscle fascicles is known as *endomysium*

(E) An intact muscle is surrounded by connective tissue called *endomysium*

181. The multinucleate arrangement of skeletal muscle during development is produced by

(A) duplication of DNA in myoblasts without cytokinesis

(B) fusion of mononucleate myoblasts

(C) cell proliferation of myotubes

(D) hypertrophy of myoblasts

(E) satellite cell differentiation

182. The transverse tubule system

(A) is part of the smooth endoplasmic reticulum

(B) depolarizes during muscle relaxation

(C) regulates the permeability of the sarcoplasmic reticulum to Ca^{2+}

(D) encircles the H band

(E) is not found in cardiac muscle

Questions 183–185

In a given muscle fiber at rest, the length of the I band is 1.0 μm and the A band is 1.5 μm.

183. What is the length of the sarcomere?

(A) 4.0 μm
(B) 3.5 μm
(C) 2.5 μm
(D) 2.0 μm
(E) 1.5 μm

184. Contraction of the muscle fiber described above results in a 10 percent shortening of the sarcomere length. What is the length of the A band after the shortening produced during muscle contraction?

(A) 1.50 μm
(B) 1.35 μm
(C) 1.00 μm
(D) 0.90 μm
(E) 0.45 μm

185. In the same muscle fiber described above, there is a 20 percent contraction of the resting length. What would be the width of the I band in the contracted muscle?

(A) 2.00 μm
(B) 1.20 μm
(C) 0.80 μm
(D) 0.50 μm
(E) 0.25 μm

186. Alterations of Ca^{2+} are essential to the normal function of skeletal muscle. In the presence of high Ca^{2+},

(A) binding sites for myosin heads on actin are blocked by the troponin-tropomyosin complex
(B) binding of Ca^{2+} to troponin C unmasks the myosin-binding site on the actin filament
(C) troponin C is inactive
(D) troponin T is responsible for the positioning of the troponins at the myosin-binding site
(E) troponin I stimulates the interaction of actin and myosin

187. Troponins C, T, and I along with tropomyosin are

(A) thin filament proteins essential for actin-myosin interaction
(B) thick filament proteins essential for actin-myosin interaction
(C) thin filament proteins essential for calcium regulation of actin-myosin interaction
(D) thick filament proteins essential for calcium regulation of actin-myosin interaction
(E) polymers of G actin

188. The sarcoplasmic reticulum of skeletal muscle is a

(A) site of a calcium-binding protein (calsequestrin) and a Ca^{2+}-activated ATPase
(B) site of glycogen storage and degradation to glucose
(C) site from which Ca^{2+} is passively transported during muscle relaxation
(D) site into which calcium is released during muscle relaxation
(E) form of rough endoplasmic reticulum

189. Observation of a histologic preparation of muscle indicates the presence of neuromuscular junctions, cross-striations, and peripherally located nuclei. The use of histochemistry demonstrates a strong staining reaction for succinic dehydrogenase. The same tissue prepared for electron microscopy demonstrates many mitochondria in rows between myofibrils and underneath the sarcolemma. The best description of this tissue is

(A) white muscle fibers
(B) fibers that contract rapidly, but are incapable of sustaining continuous heavy work
(C) red muscle fibers
(D) cardiac muscle
(E) smooth muscle

190. Which of the following is a characteristic of smooth muscle contraction?

(A) Smooth muscle is the most specialized type of muscle in terms of the contractile mechanism
(B) A sarcoplasmic reticulum as extensive as that in skeletal muscle is present in the cytoplasm
(C) T tubules are present throughout the smooth muscle cytoplasm
(D) Smooth muscle contains a regular pattern of diads of one cisterna and one T tubule for Ca^{2+} release and sequestration
(E) Contractility of smooth muscle is similar to motility in nonmuscle cells

DIRECTIONS: Each numbered question or incomplete statement below is
NEGATIVELY phrased. Select the **one best** lettered response.

191. All the following statements
are true of muscle EXCEPT

(A) smooth muscle contracts slowly
compared with skeletal muscle
and is under involuntary control
(B) muscle cells increase in diameter by adding new myofibrils, a
process called *hypertrophy*
(C) muscle contraction is dependent
on the cytoskeleton
(D) smooth muscle is multinucleate
(E) differentiated smooth muscle
cells retain the ability to undergo hyperplasia

192. All the following statements
are true of contraction of skeletal
muscle EXCEPT

(A) in the resting state, ATP hydrolysis does not occur on the
myosin heads and there is no
binding to actin filaments
(B) tight binding of myosin to actin
is dependent upon the release of
inorganic phosphate (P$_i$) from
myosin heads
(C) the binding of myosin to actin
induces the molecular folding of
myosin that results in the walking of the myosin heads along
actin filaments
(D) recycling and resetting of the
mechanism occurs through the
release of ADP from myosin
heads
(E) rigor results from the absence of
ATP

193. All the following are subunits
of myosin EXCEPT

(A) heavy meromyosin
(B) light meromyosin
(C) S1 and S2 segments
(D) P light chain
(E) troponin

194. In Duchenne muscular dystrophy, the actin-binding protein
dystrophin is absent or defective.
Dystrophin contains similar actin-
binding domains to the spectrins
(I and II) and α-actinin and is believed to have a similar function. All
the following may occur with
Duchenne muscular dystrophy
EXCEPT

(A) abnormal running, jumping, and
hopping
(B) muscle weakness
(C) deficiency in actin-binding to
skeletal muscle membrane
(D) respiratory failure
(E) deficiency in the synthesis of
actin

195. The actin-rich cell cortex is involved in all the following cell functions EXCEPT

(A) maintenance of mechanical
strength
(B) phagocytosis
(C) cytokinesis
(D) cell locomotion
(E) bidirectional transport of vesi-
cles

196. All the following statements are true of cardiac muscle EXCEPT

(A) it is capable of extensive regeneration
(B) intercellular junctions include desmosomes
(C) intercellular junctions include gap junctions
(D) intercellular junctions include fascia adherens
(E) atrial cardiomyocytes function as endocrine cells

197. In smooth muscle each of the following plays a role in regulation of contraction EXCEPT

(A) calmodulin
(B) calcium
(C) troponin
(D) myosin light chain kinase
(E) actin and tropomyosin in an arrangement similar to that of skeletal muscle

DIRECTIONS: Each group of questions below consists of lettered headings followed by a set of numbered items. For each numbered item select the **one** lettered heading with which it is **most** closely associated. Each lettered heading may be used **once, more than once, or not at all.**

Questions 198–202

Match each description with the correct labeled structure in the accompanying electron micrograph from striated muscle.

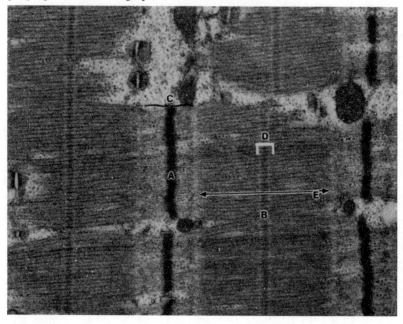

198. The sarcomere is defined as the distance between two of these structures

199. Thin filaments are anchored to this structure

200. This structure bisects the H band and is formed predominantly of creatine kinase

201. No overlap of thick and thin filaments occurs in this zone

202. This portion of the A band consists solely of the rodlike portions of myosin

Questions 203–207

Match each description with the appropriate substance

(A) Myosin (myosin II)   (F) Dynamin
(B) Minimyosin (myosin I)   (G) Gelsolin
(C) Dynein   (H) Tropomyosin
(D) Kinesin   (I) Villin
(E) Filamin   (J) Spectrin

203. A mechanochemical enzyme that mediates ciliary and flagellar bending

204. A mechanochemical enzyme that can be found on the surfaces of cellular organelles where it mediates movement toward the plus end of microtubules

205. A Ca^{2+}-binding protein that fragments actin networks in localized cytoplasmic areas and counteracts filamin

206. A mechanochemical enzyme that consists of a single globular head that binds to organelles and propels them along actin filaments

207. A mechanochemical enzyme that mediates sliding between adjacent cytoplasmic microtubules

Muscle and Cell Motility

Answers

180. The answer is B. *(Junqueira, 8/e, pp 181–182. Ross, 3/e, p 215. Stevens, pp 226–227.)* Skeletal muscle is characterized by the presence of multinucleate myofibers with striations due to the arrangement of myofilaments within the myofibrils within the cells. Nuclei are peripherally placed within the myofiber. Muscle is surrounded by connective tissue. The epimysium surrounds the entire muscle, the perimysium surrounds each bundle of muscle fibers (fascicles), and the endomysium surrounds each myofiber. The function of these connective tissues is bundling as well as transmission of blood vessels and nerves. Individual skeletal muscle cells (fibers) are elongate and long, but do not extend the entire length of the muscle. At the end of the muscle there are specialized connective tissue structures that transmit the force generated by muscle fiber contraction to bone (tendons) or to sheets of connective tissue between muscle (fascia).

181. The answer is B. *(Alberts, 3/e, pp 1176–1178. Gilbert, 4/e, pp 327–330. Junqueira, 8/e, p 183.)* In the development of skeletal muscle, myoblasts of mesodermal origin undergo cell proliferation. Myoblasts, which are mononucleate cells, fuse with each other end to end to form myotubes. This process requires cell recognition between myoblasts, alignment, and subsequent fusion. The multinucleate organization is derived from the fusion process and not by amitosis (failure of cytokinesis after DNA synthesis). This has been verified by experimental studies using allophenic mice formed by the fusion of embryos from two strains of mice. Each strain is chosen because it synthesizes a different form of a marker enzyme. The presence of hybrid forms in skeletal muscle myotubes, but not other tissues of the mouse, verifies the fusion process in myotube formation. Mitotic activity is terminated after fusion occurs. Fibroblast growth factor (FGF) is an important regulator of these processes since it stimulates myoblast cell proliferation and inhibits differentiation into myotubes. Satellite cells are an adult stem cell population of myoblasts that persist from fetal life. With an appropriate stimulus, they can proliferate and subsequently differentiate into new muscle cells.

182. The answer is C. *(Alberts, 3/e, pp 854–855. Junqueira, 8/e, pp 184–186. Ross, 3/e, pp 220–223. Stevens, p 61. Widnell, pp 178–179.)* The transverse tubule system, or T system, is an extension of the plasma membrane of the myofiber (sarcolemma). In combination with the paired terminal cisternae, the transverse tubules form a triad. Two triads are found in each sarcomere of skeletal muscle—one at each junction of dark (A) and light (I) bands. This positioning results in the close association of the transverse tubule system with each myofibril. Depolarization of the T system during contraction is transmitted to the sarcoplasmic reticulum at the triad. The sarcoplasmic reticulum contains many calcium channels, which open in response to depolarization and result in a massive increase in cytosolic calcium. During repolarization of the T system, Ca^{2+} is transported from the cytosol and sequestered in the sarcoplasmic reticulum through the activity of Ca^{2+}-ATPase and calsequestrin, respectively. The T-tubule system thereby provides a mechanism for the regulation of intracellular calcium, and the troponin-tropomyosin complex regulates the interaction of actin and myosin through the binding of calcium by troponin C and the subsequent exposure of the myosin-head binding site on actin. Calcium is responsible for the coupling of excitation and contraction in skeletal muscle. Cardiac muscle also has a T system, although it is not as elaborate and well organized as that found in skeletal muscle (e.g., diads are present rather than the triads of skeletal muscle and there are fewer T tubules in the atrial versus ventricular muscle).

183–185.   The answers are 183-C, 184-A, 185-D. *(Alberts, 3/e, pp 847–849. Junqueira, 8/e, pp 185–186. Ross, 3/e, pp 218–221. Stevens, pp 58–59. Widnell, pp 171–172.)* During contraction the sarcomere, the distance between adjacent Z lines, decreases in length and the length of the A band is almost constant. However, as the degree of overlap of thick and thin filaments is altered, the thin filaments, which form the I band and are anchored to the Z line, are pulled toward the center of the sarcomere. As this occurs, the I band decreases in length and the H band is no longer visible. The filaments themselves do not decrease in length—they slide past one another in the sliding-filament model of muscle contraction.

The average length of a sarcomere is 2.5 μm. This distance is measured from one Z line to the next Z line. If the resting length of the A band is 1.5 μm and the length of the I band is 1.0 μm, then the resting length of the sarcomere is determined by adding the length of the I band to the length of the A band. If there is a 20 percent contraction of the muscle (contraction to 80 percent of its length), then the sarcomere is reduced in length from 2.5 to 2.0 μm. The size of the A band remains unchanged (whether the contraction is 10 or 20 percent); therefore the length of the I band is reduced from 1.0 to 0.5 μm and makes up for the 0.5 μm reduction in length during muscle contraction.

186.   The answer is B. *(Alberts, 3/e, pp 854–855. Junqueira, 8/e, pp 185–186. Stevens, p 60. Widnell, pp 177–178.)* The proteins troponin and tropomyosin regulate the interaction of actin and myosin through their response to altered concentrations of calcium. In the resting state, as defined by low calcium levels (generally below 10^{-8} M), the conformation of the troponin-tropomyosin complex blocks the myosin-binding site. When calcium levels are elevated (10^{-6} to 10^{-5} M), the myosin-binding site is exposed. This occurs through the binding of calcium to troponin C (named for its calcium-binding activity). Troponin T (named for its binding to tropomyosin) is responsible for binding of the troponin complex to tropomyosin and the positioning of the troponins at the myosin-binding site during low calcium conditions. Troponin I (named for its inhibitory function) interferes with the binding of myosin heads to actin. Troponin C is a calmodulin-like molecule required for the calcium dependency of this response.

187. The answer is C. *(Alberts, 3/e, pp 854–855. Junqueira, 8/e, pp 182– 190. Stevens, p 60. Widnell, pp 177–178.)* The actin filament is composed of polymerized G actin, which forms the filamentous F-actin strand. Associated with actin are other thin filament proteins called the *troponins* and *tropomyosin*, which are essential for calcium regulation of the actin-myosin interaction. These regulatory proteins are not required for interaction of actin and myosin, which will occur independently of calcium concentration. If the troponins and tropomyosin are added to an in vitro mixture of actin and myosin, then the interaction of actin and myosin becomes calcium-dependent.

188. The answer is A. *(Alberts, 3/e, pp 853–855. Junqueira, 8/e, pp 184– 185. Stevens, p 61. Widnell, pp 178–179.)* The sarcoplasmic reticulum provides a mechanism for the muscle cell to regulate the concentration of cytosolic calcium. It is a modified smooth endoplasmic reticulum that serves alternatively as a storage site and a source of cellular calcium. Calcium is actively transported from the cytosol to the sarcoplasmic reticulum through the activity of a Ca^{2+}-dependent ATPase. Calsequestrin is a calcium-binding protein found in the sarcoplasmic reticulum. From its name it is clear that it functions in the sequestration of calcium within the sarcoplasmic reticulum. Calcium is released from the sarcoplasmic reticulum during muscle contraction; it is stored during relaxation. Glycogen is stored as particles or droplets in the cytoplasm, which contains the enzymes required for glycogen's synthesis and breakdown.

189. The answer is C. *(Junqueira, 8/e, pp 189–190. Ross, 3/e, pp 215– 216.)* The histologic sample must be skeletal or cardiac muscle because of the presence of cross-striations. The presence of peripherally placed nuclei eliminates cardiac muscle as a possible tissue. Skeletal muscle may be subclassified into three muscle fiber types. Red muscle fibers have a high content of cytochrome and myoglobin (an oxygen storage pigment, analogous to hemoglobin found in red blood cells). Red muscle contains many mitochondria beneath the myofiber cell membrane that function in the high metabolism of these cells. Mitochondria are also found in a longitudinal array surrounding the myofibrils. The presence of numerous mitochondria provides a strong staining reaction with the use of cytochemical stains such as that for succinic dehydrogenase. Physiologically, red fibers are capable of continuous contraction but are incapable of rapid contraction. White muscle fibers would stain very lightly for succinic dehydrogenase and there would be few mitochondria visible at the ultrastructural level. These would be primarily associated with the triads as occurs in all forms of skeletal muscle. White fibers are capable of rapid contraction but are unable to sustain continuous heavy work. They are larger than red fibers and have more prominent innervation and synaptic vesicles. Human skeletal muscle fibers are composed of red, white, and intermediate type fibers. The intermediate fibers possess characteristics of both red and white fibers including a size and innervation pattern intermediate between red and white muscle fibers.

190. The answer is E. *(Alberts, 3/e, pp 856–857. Junqueira, 8/e, pp 197–200. Ross, 3/e, pp 230–234. Stevens, pp 62–65. Widnell, pp 179–181.)* Smooth muscle is the least specialized type of muscle. The contractile process is similar to the actin-myosin interactions that occur in motility of nonmuscle cells. In the smooth muscle cell, actin and myosin are located in a crisscross arrangement across the cell. Attachment occurs at dense bodies in the sarcolemma and cytoplasm to the supporting network of intermediate filaments (e.g., desmin in most smooth muscle and vimentin plus desmin in vascular smooth muscle). Both the cytoplasmic and membrane dense bodies contain α-actinin and therefore resemble the Z lines of skeletal muscle. Contraction causes cell shortening and a change in shape from elongate to more globular. The contraction process involves actin and myosin and occurs by a sliding filament action analogous to the mechanism used by thick and thin filaments in striated muscle. The connections to the plasma membrane allow all the smooth muscle cells in the same region to act as a functional unit. Sarcoplasmic reticulum is not as well developed as that in the striated muscles. There are no T tubules present; however, endocytic vesicles called *caveolae* are believed to function in a fashion similar to the T-tubule system of skeletal muscle. Cardiac muscle contains diads involved in Ca^{2+} release and sequestration.

191. The answer is D. *(Alberts, 3/e, p 847. Junqueira, 8/e, p 181.)* Muscle cells are responsible for coordinated movement within the body. Skeletal muscle is predominantly under voluntary control and is rapid in response in comparison with smooth muscle, which responds slowly and is predominantly under involuntary control. Growth of muscle may occur by mitosis (hyperplasia) or by increase in diameter by addition of new myofibrils (hypertrophy). However, in the adult only differentiated smooth muscle is capable of hyperplasia. In hypertension, the smooth muscle of blood vessels undergoes hyperplasia as part of the thickening of the arterial vessels. Skeletal and cardiac muscle in the adult undergo hypertrophy. The processes of muscle contraction and cellular motility are dependent upon the cellular cytoskeleton including actin, intermediate filaments, and microtubules. Skeletal muscle is the only type of muscle that is multinucleate.

192. The answer is A. *(Alberts, 3/e, pp 851–853. Junqueira, 8/e, pp 185–190. Stevens, p 60. Widnell, pp 175–176.)* The cycle of ATP-actin-myosin interactions during contraction begins with the resting state. In the quiescent period, ATP binds to myosin heads; however, hydrolysis occurs slowly and only allows the weak binding of myosin heads to the actin filaments. Tight binding occurs only when P_i is released from myosin heads. The tight binding of actin and myosin induces a conformational change in the myosin head, which pulls against the actin filament to cause the "powerstroke" of the myosin head's walking along the actin filament. This walking process is unidirectional and is based on the polarity of the actin filament (i.e., walking occurs from the negative to the positive end of the actin filament). Recycling occurs through the release of ADP and the subsequent addition of an ATP molecule and detachment of the myosin head from actin. Rigor results from the lack of ATP because one ATP molecule is required for each myosin molecule present in the muscle. Rigor mortis occurs from the total absence of ATP.

193. The answer is E. *(Junqueira, 8/e, p 189. Stevens, p 59. Widnell, pp 172–175.)* Myosin is composed of two coiled heavy chains and four light chains. It may be separated into heavy and light meromyosin by enzymatic treatment. Heavy meromyosin has two segments: S1 (the globular head region) and S2. The S1 subfragment includes the light chains that are associated with the globular head regions. This region is significant because it is the site of the actin binding that activates ATPase activity. S2 is a dimeric portion of the myosin molecule that connects the two S1 segments to the coiled light meromyosin subunit. The P light chain is one of the two light chains associated with the globular heads and is phosphorylated by myosin light chain kinase. In skeletal muscle, phosphorylation of the light chain is not required for binding to actin. Troponin is associated with actin, not myosin.

194. The answer is E. *(Alberts, 3/e, p 855. Isselbacher, 13/e, pp 2383–2384.)* Muscular dystrophy refers to a group of progressive hereditary disorders that involve mutations in the dystrophin gene. Dystrophin is an actin-binding protein with a structure common to spectrins I and II and α-actinin. A deficiency or absence of dystrophin has been implicated in Duchenne muscular dystrophy. Dystrophin, like these other actin-binding proteins, functions to bind actin to the skeletal muscle membrane. The inability to bind actin to the plasma membrane of skeletal muscle leads to disruption of the contraction process, weakness of muscle, and abnormal running, hopping, and jumping. Gowers' maneuver is the method used by persons suffering from muscular dystrophy to stand from a sitting position. Respiratory failure occurs in these persons because of disruption of diaphragmmatic function. Synthesis of actin is not reduced in skeletal muscle from these patients; in fact, hypertrophy and pseudohypertrophy (replacement of muscle with connective tissue and fat) occurs.

195. The answer is E. *(Alberts, 3/e, pp 813–814, 834. Goodman, pp 83, 91–92.)* The cell cortex is an area of the cell immediately underneath the plasma membrane and is rich in actin. This region is important in maintaining the mechanical strength of the cytoplasm of the cell. It is also essential for cellular functions that require surface motility. These functions include phagocytosis and cell locomotion. While movement of vesicles along filaments is regulated by minimyosins (myosin I), movement of vesicles and organelles is predominantly a function of microtubules under the influence of the bidirectional motors kinesin and dynein.

196. The answer is A. *(Junqueira, 8/e, pp 191–197. Ross, 3/e, pp 226–229. Stevens, p 62.)* Cardiac muscle is composed of mononuclear cardiomyocytes. Cardiomyocytes are joined by intercalated disks that possess desmosomes (macula adherens), fascia adherens, and gap junctions. These junctional complexes function in a fashion similar to junctional complexes between other cells in the body. The fascia adherens is found only in the intercalated disk. Unlike the zonula adherens of epithelial cells, it is not continuous around the periphery of the cell-cell junction. It does subserve a similar function in the linking of the actin filaments of each cell to the sarcolemma and the anchoring of the contractile elements of each cardiomyocyte to the adjacent cardiomyocyte. The desmosomes are involved in the linkage of intermediate filaments from adjacent cardiomyocytes. Gap junctions are responsible for communication between neighboring cells and the synchronization of contraction required for simultaneous activity through electronic coupling of atria and ventricles. Skeletal muscle contains satellite cells that may serve as stem cells for the production of new myoblasts and the differentiation of new skeletal muscle fibers. There is an absence of satellite cells in cardiac muscle, which is incapable of this form of regeneration.

The heart functions as an endocrine organ through the production of a hormone involved in fluid and electrolyte homeostasis. Atrial natriuretic peptide (ANP), which is also known as *atriopeptin* and *atrial natriuretic factor,* is a hormone synthesized and stored in the cardiac atria. It is released following atrial stretch and distention and it induces the kidneys to increase natriuresis and diuresis. The action of ANP is in opposition to aldosterone and antidiuretic hormone, which conserve sodium and fluid. ANP has been localized in extraatrial sites within the body including the ventricles of the heart, kidneys, adrenal glands, salivary glands, and in the central nervous system.

197. The answer is C. *(Alberts, 3/e, pp 856–857. Junqueira, 8/e, p 198. Stevens, pp 62–65. Widnell, pp 179–180.)* When intracellular calcium levels increase, the calcium is bound to the calcium-binding protein calmodulin. Ca^{2+}-calmodulin is required and is bound to myosin light chain kinase to form a Ca^{2+}-calmodulin-kinase complex. This complex catalyzes the phosphorylation of one of the two myosin light chains on the myosin heads. This phosphorylation allows the binding of actin to myosin. A specific phosphatase dephosphorylates the myosin light chain, which returns the actin and myosin to the inactive, resting state. The actin-tropomyosin interactions are similar in smooth and skeletal muscle; however, troponin is absent in smooth muscle.

198–202. The answers are 198-A, 199-A, 200-B, 201-C, 202-D. *(Junqueira, 8/e, pp 181–185. Ross, 3/e, pp 217–220. Stevens, pp 58–59. Widnell, pp 171–172.)* Myofibrils are composed of sarcomeres, which are repeating units that extend from Z line to Z line. With the use of polarizing microscopy the A (anisotropic) bands are visible as dark, birefringent structures and the I (isotropic) bands are visible as light-staining bands. The I band consists of thin filaments without overlap of thick filaments. At the center of the myofibril and consisting of thick filaments is the A band, which interdigitates with the I band. Each I band is bisected by the Z line. The Z line is composed mostly of the intermediate filament protein desmin and other proteins such as α-actinin, filamin, and amorphin, as well as Z protein. In the center of the A band is a lighter staining area that consists only of thick (rodlike portions of myosin) filaments and is known as the *H band*. Lateral connections occur between adjacent thick filaments in the region of the M line, which bisects the H zone and is composed primarily of creatine kinase, an enzyme that catalyzes the formation of ATP.

203–207. The answers are 203-C, 204-D, 205-G, 206-B, 207-F. *Alberts, 3/e, pp 813–814, 816–818, 835–839, 843–845. Stevens, p 20. Widnell, pp 183–188.)* Myosin, minimyosin, dynein, dynamin, and kinesin are all mechanochemical enzymes or molecular "motors" that hydrolyze ATP and undergo conformational changes that are converted into movement of cells or organelles. Kinesin moves vesicles unidirectionally from the minus end to the plus end of the microtubule, e.g., from the cell body to the axon terminus in fast axonal transport. Cytoplasmic dynein is responsible for movement ɛ microtubules in the opposite direction (i.e., toward the minus end of the crotubules). Ciliary and flagellar bending is the classic model for microtubule-based motility. The motor is dynein, which causes the relative sliding between microtubules in the axoneme. Dynein arms extend from the A subfiber of one doublet to the B subfiber of the adjacent doublet in the axoneme. Cyclic formation of dynein cross-bridges between adjacent doublets causes relative sliding of the doublets. Structural constraints within the axoneme as a whole convert sliding into ciliary bending. Binding to organelles propels them along actin filaments. Dynamin is a relatively newly discovered ATPase motor that mediates sliding between adjacent cytoplasmic microtubules.

Actin-binding proteins are essential to the normal function of the cytoskeleton and organize different arrangements of actin filaments within the cytoplasm. There are a number of actin-binding proteins with a variety of functions: stabilizing, sliding, bundling, cross-linking, fragmenting, capping, contraction, and movement of vesicles. Mechanochemical enzymes such as myosin I and myosin II are also classified as actin-binding proteins. Minimyosins (myosin I) have a single globular myosin head that binds to membranes and membrane-bound organelles. Myosin (myosin II) is a coiled molecule consisting of bipolar filaments. It hydrolyzes ATP to ADP plus P_i when stimulated by actin binding. Myosin II is a family of myosins, all of which have two heads and a rodlike tail. The heads have ATPase activity and motor activity for sliding filaments as occurs in muscle.

Tropomyosin is a rod-shaped, rigid molecule that binds along the actin filaments, resulting in stiffening and stabilization. Villin is an actin-binding protein present in microvilli and is essential for the formation and maintenance of the microvillus structure prevalent in some types of epithelium (e.g., the epithelium of the small bowel).

Filamin or other actin cross-linking proteins form a gel network in the cell cortex (the area just beneath the cell membrane). The presence of the actin gel in the cell cortex contributes to the rigidity of the cell and is also involved in changes in cell shape. Gelsolin is a Ca^{2+}-binding protein that fragments actin networks in localized cytoplasmic areas. This is a prerequisite for cell movement. Gel-sol transformations in which the actin network becomes liquefied are controlled by Ca^{2+} in the presence of the actin-binding protein gelsolin. In the gel state, actin may be bound by filamin or another actin-binding protein, which prevents contraction. The addition of calcium or gelsolin breaks the filamin link and allows contraction if myosin is present. Spectrin cross-links filaments, by their sides, to the plasma membrane.

Nervous System

DIRECTIONS: Each question below contains five suggested responses. Select the **one best** response to each question.

208. In general, neurons have

(A) a small nucleolus
(B) a paucity of rough endoplasmic reticulum (RER)
(C) a highly ordered cytoskeleton to provide internal support
(D) a poorly organized microtubule network
(E) few lysosomal elements

209. In the histogenesis of the neural tube, which zone will become the white matter of the adult CNS?

(A) Ventricular zone
(B) Marginal zone
(C) Mantle zone
(D) Ependymal zone
(E) Intermediate zone

210. In the histogenesis of the cerebral cortex, the layers are formed

(A) as waves of proliferation from an external granular layer
(B) randomly as cells are born
(C) in a sequence from deep to superficial
(D) to establish an area of adult peripheral white matter
(E) to establish an identical pattern in each area of the adult cortex

211. Which of the following statements is true of cerebellar histogenesis?

(A) Two germinative (proliferative) layers are established during development
(B) All gray matter forms by cell birth from the ventricular zone of the neural tube
(C) The layers of the cerebellar cortex are formed by successive waves of cell birth in the ventricular zone
(D) Astroglia and oligodendroglia form from infiltrating cells
(E) The molecular layer is formed by Purkinje cells

212. The cells responsible for the entry of human immunodeficiency virus (HIV) into the CNS are

(A) microglia/macrophages
(B) astrocytes (astroglia)
(C) oligodendrocytes (oligodendroglia)
(D) endothelial cells
(E) Schwann cells

213. The neuroepithelium of the developing neural tube gives rise to

(A) oligodendrocytes
(B) sympathetic chain ganglia
(C) dorsal root ganglia
(D) microglia
(E) sensory neurons

214. Which of the following statements is true of the histologic structure of neurons?

(A) Nissl is present in the axons of motor neurons
(B) Pseudounipolar neurons are found predominantly in the retinal and olfactory epithelia and in the vestibular and acoustic ganglia
(C) Bipolar neurons are found predominantly in the dorsal root ganglia
(D) The dendrites and perikarya are receptive to environmental stimuli
(E) Axons are multiple, short processes that contain extensive rough endoplasmic reticulum

215. Which of the following statements is true of the action potential?

(A) It results from hyperpolarization
(B) It results directly from the opening of K^+ channels
(C) It is an all-or-none phenomenon
(D) It is of variable amplitude and duration for a given axon
(E) Conduction rate is independent of myelination

216. Which of the following occurs in axonal transport?

(A) Mitochondria and secretory vesicles are transported in an anterograde direction by slow axonal transport
(B) Actin and metabolic enzymes are transported by slow retrograde transport
(C) Receptors and recycled membranes from the synapse are transported down the axon by slow axonal transport
(D) Dynein is used as a microtubule motor for slow axonal transport
(E) Endocytosis followed by rapid retrograde transport can provide a route for toxins or viruses to reach the perikarya

217. In comparing and contrasting myelination in the central and peripheral nervous systems, which of the following is true?

(A) Schwann cells myelinate several adjacent axons
(B) Myelin formed in the two locations is identical
(C) Myelin increases membrane capacitance
(D) All myelin in the CNS develops prenatally
(E) The myelin-forming cell cytoplasm contributes to the myelin sheath

218. The central nervous system component illustrated in the light micrograph below

(A) is a cross-section of the spinal cord
(B) contains pyramidal cells that in different layers of the cortex project to different parts of the nervous system
(C) contains a central area of gray matter
(D) contains vertical columns that are perpendicular to the surface and subserve one sensory modality
(E) contains modified astrocytes called *Bergmann glia* and basket cell axons situated in the vicinity of Purkinje cells

219. The neurons shown in the accompanying photomicrograph of a peripheral ganglion have which of the following functions?

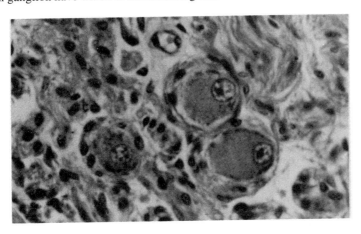

(A) Autonomic transmission
(B) Conduction of pain impulses
(C) Proprioception
(D) Innervation of striated muscle
(E) Transmission of visual stimuli

---

220. Regeneration of axons

(A) occurs in the segment distal to the damage
(B) is independent of the survival of the perikaryon
(C) includes a decrease in the volume of the perikaryon
(D) is dependent on proliferation of Schwann cells
(E) is initiated with an increase in production of Nissl

221. The nodes of Ranvier

(A) occur only in the CNS
(B) contain few Na^+ gated channels
(C) represent the midpoints of myelination segments
(D) are completely covered by myelin
(E) increase the efficiency of nerve conduction

222. Which of the following directly contributes to the blood-brain barrier?

(A) Fenestrations between brain capillary endothelial cells
(B) Desmosomes between brain capillary endothelial cells
(C) The glia limitans, which surrounds the CNS
(D) Neuron-oligodendrocyte interactions
(E) Microglial function

223. At the neuromuscular junction, action potentials are coupled to neurotransmitter release by voltage-gated

(A) Ca^{2+} channels
(B) Na^+ channels
(C) K^+ channels
(D) Cl^- channels
(E) gap junctions between the presynaptic terminal and the muscle cell

224. In the photomicrograph below of tissue taken from the cerebral cortex, the cell illustrated is a

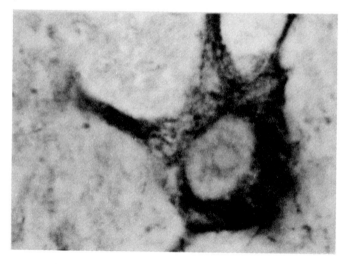

(A) basket cell
(B) granule cell
(C) neuroglial cell
(D) pyramidal cell
(E) Purkinje cell

DIRECTIONS: Each numbered question or incomplete statement below is NEGATIVELY phrased. Select the **one best** lettered response.

225. In the photomicrograph in the previous question, all the following structures are demonstrable EXCEPT the

(A) axon
(B) dendrites
(C) euchromatin
(D) Nissl bodies
(E) nucleolus

226. The neural crest gives rise to all the following EXCEPT

(A) adrenal medulla
(B) Schwann cells
(C) cells that compose the pia and arachnoid of the meninges
(D) ventral horn cells
(E) sensory cells of the cranial ganglia

227. All the following statements are true of the cerebrospinal fluid (CSF) EXCEPT

(A) CSF is produced by the choroid plexuses
(B) CSF is secreted by an active process into the subarachnoid space
(C) CSF can be safely sampled by a lumbar puncture at the L1 level
(D) CSF is absorbed by the venous sinuses through the arachnoid villi
(E) hydrocephalus may be caused by either blockage of the median and lateral apertures of the fourth ventricle or a decrease in CSF absorption

228. All the following are components of sensory ganglia EXCEPT

(A) axons
(B) synapses
(C) perikarya
(D) loose connective tissue
(E) satellite or Schwann cells

229. All the following are characteristics or functions of astrocytes EXCEPT that they

(A) are derived from a bone-marrow monocyte lineage
(B) stain positively with glial fibrillary acidic protein (GFAP)
(C) assist in the formation of the blood-brain barrier
(D) form the scaffolding for guidance of developing neurons
(E) form the glia limitans

DIRECTIONS: Each group of questions below consists of lettered headings followed by a set of numbered items. For each numbered item select the **one** lettered heading with which it is **most** closely associated. Each lettered heading may be used **once, more than once, or not at all.**

Questions 230–233

Match each structure with the appropriate label on the electron micrograph. If no label applies, answer E for "none of the above."

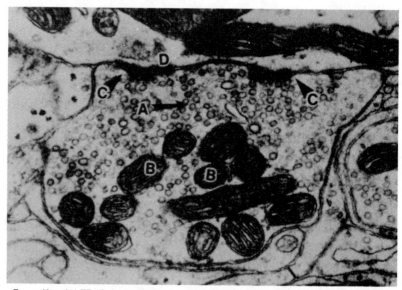

From Kandel ER, Schwartz JH, Jessell TM: *Principles of Neural Science,* 3/e. Elsevier, 1991, with permission. Courtesy of Drs. JE Heuser and TS Reese.

230. Mitochondria

231. Synaptic vesicles

232. Postsynaptic membrane

233. Presynaptic membrane

Questions 234–238

Match each description with the appropriate cell.

(A) Microglia
(B) Astrocytes
(C) Oligodendrocytes
(D) Schwann cells
(E) Ependymal cells
(F) Satellite cells

234. Antigen-presenting cells that stain positively with antisera for common leukocyte antigen and class II major histocompatibility complex (MHC)

235. Prominent desmosomal junctions with each other and apical regions with microvilli and cilia

236. Production of myelin in the CNS

237. Responsibility for scar formation during damage repair, a process called *gliosis*

238. Derivation of the most common type of glioma

Questions 239–243

Match each description with the layer seen between the skull and the CNS.

(A) Periosteum
(B) Epidural space
(C) Dura mater
(D) Subdural space
(E) Arachnoid
(F) Subarachnoid space
(G) Pia mater

239. Primary location of the cerebrospinal fluid

240. Location of the venous sinuses

241. Lining of the perivascular spaces

242. Retention of the most osteogenic potential in the adult

243. Intermediate layer of the meninges

Nervous System

Answers

208. The answer is C. *(Junqueira, 8/e, pp 155–157. Ross, 3/e, pp 256–260. Stevens, pp 206–207.)* In general, neurons are metabolically active cells that carry out considerable transcription of mRNA from DNA, translation of mRNA, and protein synthesis. One would therefore expect to see a large nucleolus and extensive RER. Neurons possess a well-organized cytoskeleton composed of microfilaments, microtubules, and intermediate filaments. In neurons, the intermediate filaments are composed of neurofilament proteins not found in the glia. These cytoplasmic components are required to internally support the cellular shape since the cell may have very long dendritic or axonal extensions. In addition the microtubules are required for transport of organelles and intracellular materials along the axon. Lysosomal components are found in large quantities in neurons because of the extensive synthesis and degradation of membranous components within these cells. With age, neurons may contain increased amounts of lipofuscin in residual bodies. These structures represent undigested organelles and other intracellular materials for which there is inadequate concentration of lysosomal enzymes.

209. The answer is B. *(Junqueira, 8/e, p 152. Larsen, p 380. Moore, Before We Are Born, 4/e, pp 275, 279–280. Moore, Developing Human, 5/e, pp 385, 389–391.)* During the differentiation of the neural tube there are two distinct stages of development. The first stage involves cell proliferation and occurs before neural tube closure. The second stage involves the differentiation of neurons from the germinal (ventricular) layer of the epithelium and is initiated after closure of the neural tube. Differentiation of three distinct layers of the wall is observed. Mitotic activity occurs in the ventricular zone, closest to the lumen. The next zone is the mantle (intermediate) zone, where cell bodies of differentiating motor neurons are located. The most peripheral zone is the marginal zone, which contains the myelinated axons of the developing motor neurons (adult white matter).

In the central nervous system there is a separation of neural tissue into two types. White matter contains a predominance of myelinated fibers, which contribute to the white appearance and name. The gray matter contains mostly cell bodies of neurons. The formation of three layers in the developing neural tube results in the pattern of peripheral white matter with a central H-shaped region of gray matter, which is seen in the spinal cord. In the case of the cortex (cerebellar or cerebral), there is secondary proliferation to form peripheral areas of gray matter.

Astrocytes and oligodendrocytes, the macroglia, arise from the neural epithelium and not from neural crest cells. After neuroblast and glial differentiation occurs, the remaining ventricular zone cells form the ependymal layer.

210. The answer is C. *(Larsen, pp 403–404. Moore, Before We Are Born, 4/e, pp 294–295. Moore, Developing Human, 5/e, p 409.)* The cerebral cortex develops in successive waves of cell proliferation from the ventricular zone and the subventricular zone, which replaces the ventricular zone as an area of cell proliferation part way through the differentiation of the cerebral cortex. The layers are formed in a sequence from deep to superficial. The transient structures formed during cerebral cortical development include the cortical plate and the subplate, both located between the intermediate (mantle) and marginal zones. Cerebral as well as cerebellar cortical histogenesis involves the development of peripheral areas of gray matter (cortex). Cerebral cortical development differs in timing and pattern (extent of layers I–VI) from region to region.

211. The answer is A. *(Larsen, pp 389–391. Moore, Before We Are Born, 4/e, pp 287–289. Moore, Developing Human, 5/e, pp 402–403.)* The cerebellum forms from the rhombic lips in that region of the embryonic rhombencephalon (hindbrain) called the *metencephalon*. It forms from the dorsal region (essentially the sensory part) of this structure called the *alar plates*. The neural tube differentiates into ventricular, mantle, and marginal zones as occurs in other regions of the developing nervous system. The ventricular zone is the site of birth of neuroblasts, which establish the cerebellar nuclei (i.e., the deep cerebellar nuclei: dentate, globose, emboliform, and fastigial nuclei), the Purkinje cells, and the Golgi cells of the cerebellum. Later in development a second proliferative zone is established in the marginal zone. This external germinative (granular) layer is the source of neuroblasts that form the basket, granule, and stellate cells. The granule cells and some of the basket cells form the granular layer of the adult (deep to the Purkinje cells), while the stellate cells and the remainder of the basket cells form the molecular layer of the adult cerebellar cortex (superficial to the Purkinje cells). Macroglia (astrocytes and oligodendrocytes) are born in both the ventricular zone (internal germinative layer) and the external germinative layer, which is established later in development. The Purkinje cells form their own definitive layer.

212. The answer is A. *(Budka, Acta Neuropathol 77:225–236, 1989. Isselbacher, 13/e, pp 1589–1591.)* Microglia/macrophages are infected with HIV and carry the virus into the CNS. The virus remains latent until a stimulus activates viral production. These cells are the most conspicuous elements of HIV-induced CNS pathology. Infection, proliferation, and fusion of microglia/ macrophages appear to be involved in the development of giant cell encephalitis of acquired immune deficiency syndrome (AIDS) and other pathologies associated with neuronal damage in AIDS dementia. The CNS effects of AIDS are extensive as indicated by the fact that 90 percent of AIDS patients show abnormalities in the cerebrospinal fluid (CSF), even in asymptomatic stages of the disease.

213. The answer is A. *(Junqueira, 8/e, p 152. Kandel, 3/e, pp 297–300. Moore, Before We Are Born, 4/e, pp 280, 282. Moore, Developing Human, 5/e, pp 390, 393.)* The neural tube is the embryonic source of the central nervous system, i.e., the brain and the spinal cord. It differentiates through an induction by the underlying notochord and the laterally positioned paraxial mesoderm. Motor and association neurons, macroglia (i.e., oligodendrocytes and astrocytes), and ependymal cells of the ventricles and choroid plexus form from the neuroectoderm of the neural tube. Other cell types are derived from the neural crest, while microglia form from mesoderm (i.e., they are bone-marrow-derived from the monocyte lineage).

214. The answer is D. *(Burkitt, 3/e, pp 112–117. Erlandsen, p 61. Junqueira, 8/e, pp 152–156. Ross, 3/e, pp 256–260. Stevens, pp 206–207.)* Neurons are excitatory cells composed of three major parts: the perikaryon (cell body), dendrites, and generally a single axon. The dendrites and the perikaryon are capable of receiving stimuli and compose the receptive domain of the neuron. The integrative domain is the perikaryon, the area of the neuron surrounding the nucleus, and the conducting domain is the axon of the neuron. Generally, the dendrites are multiple and short, while there is a single axon that is usually longer than the dendrites.

Neurons are found in several classifications: pseudounipolar (one single process, which branches into central and peripheral processes after leaving the perikaryon, e.g., perikarya of the dorsal root ganglia), bipolar (one axon and one dendrite, e.g., the cells of the retinal and the olfactory epithelia and the vestibular and acoustic ganglia), and multipolar (e.g., ventral horn cells).

Nissl is found in the perikarya and in the dendrites of all neurons. Nissl can be identified easily using a toluidine blue stain, and it represents the presence of extensive RER, including extensive numbers of ribosomes. It is absent from the axons and may be used as a method of distinguishing the axons from the dendrites.

215. The answer is C. *(Alberts, 3/e, p 531. Junqueira, 8/e, pp 158–160. Kandel, 3/e, pp 28–29, 104–108, 111–112. Stevens, p 208.)* In the resting state, sodium and potassium pumps build up high ionic gradients across the axolemma. The action potential in a neuron results from depolarization following the opening of Na^+ channels. The resting potential is about -90 mV and is displaced toward 0 volts. When the threshold voltage is reached, sodium channels open and sodium ions enter the neuron. K^+ channels also open in response to changes in membrane potential but bring the membrane to a hyperpolarized state. The action of inward flux of K^+ combined with the closing of Na^+ channels is important in the return to the resting membrane potential. The action potential is an all-or-none phenomenon and occurs with constant amplitude and duration for a given axon. Myelination results in a much more rapid conduction of the action potential.

216. The answer is E. *(Alberts, 3/e, p 814. Junqueira, 8/e, p 157. Kandel, 3/e, pp 57–61. Stevens, p 28.)* Axonal transport occurs by several different mechanisms. Slow axonal transport involves the transport of cytoskeletal elements such as actin, tubulin, and neurofilaments from the perikaryon down the axon. Slow transport occurs at a velocity of 1 to 5 mm/day. Dendritic transport occurs in a manner similar to that of slow axonal transport. In contrast to slow axonal transport, rapid anterograde (away from the perikaryon) transport and retrograde (toward the perikaryon) transport occur at rates of 200 to 300 mm/day. Membrane-bound organelles such as newly formed secretory vesicles and mitochondria are transported rapidly in an orthograde or anterograde direction. Receptors, recycled membranes, and worn-out organelles are transported following a retrograde mechanism back to the cell body of the neuron. This mechanism may also be used by extracellular materials, which may be endocytosed by a receptor-mediated process and subsequently transported to the perikarya by rapid retrograde transport. Such a pathway may be followed in the CNS by toxins such as tetanus toxin and viruses. Retrograde transport is also used experimentally by neuroanatomists to map connections in the CNS in animals.

Colchicine and other microtubule toxins block fast axonal transport, which indicates a dependence on microtubule-based movements for axonal transport. The molecular motors for rapid retrograde and anterograde transport differ. Kinesin is a microtubule motor that hydrolyzes ATP and in so doing moves organelles along microtubules in an anterograde direction (toward the plus ends of the microtubules). Dynein is also an ATPase and a microtubule-based motor that moves organelles along microtubules in a retrograde direction (toward the minus ends of the microtubules).

217. The answer is E. *(Burkitt, 3/e, pp 117–121. Junqueira, 8/e, pp 169–170. Kandel, 3/e, pp 43–45. Ross, 3/e, pp 264–267, 271. Stevens, pp 210–212.)* Myelination in the central (CNS) and peripheral (PNS) nervous systems occurs by similar methods, although there are differences in the supportive cells responsible for myelination. In the CNS, the oligodendrocytes myelinate axons while the Schwann cells carry out myelination in the PNS. Oligodendrocytes myelinate several axons at one time while the Schwann cells myelinate only one axon at a time. Myelin is similar in both locations but differs in the presence of Schmidt-Lanterman clefts, which only appear in the PNS and represent the presence of Schwann cell cytoplasm that is not displaced toward the periphery during the formation of myelin. Myelin is an insulator and also decreases membrane capacitance. White matter is high in myelin content and is named by the presence of tracts of axons that appear white (myelinated). Gray matter represents neuron-rich areas low in myelin (e.g., cell bodies).

In the PNS formation of myelin is initiated by the invagination of an axon into a Schwann cell. A mesaxon is formed as the outer leaflets of the cell membrane fuse. Subsequently, the mesaxon of the Schwann cell wraps itself around the fiber. In the CNS, oligodendrocytes form myelin around several axon segments as compared with the 1:1 relationship between Schwann cells and axon segments in the PNS. Myelination occurs in both pre- and postnatal development. For example, the extrapyramidal system is myelinated postnatally and its maturation may be assessed by use of the Babinski reflex. CNS myelin is the target for attack by components of the immune system in multiple sclerosis.

218. The answer is E. *(Burkitt, 3/e, p 370. Junqueira, 8/e, pp 164–167. Kandel, 3/e, p 630. Ross, 3/e, pp 281–282, 296–299. Stevens, pp 216, 223–224.)* The light micrograph illustrates the histologic structure of the cerebellar cortex. This structure contains three layers—molecular, Purkinje cell, and an inner granular layer—and a central core of white matter in which the nuclei of the gray matter are located. The cerebellum also contains special astrocytes called *Bergmann glia* and basket cell axons, which are located close to Purkinje cells.

The cerebral cortex consists of six layers. Layers II, III, V, and VI contain pyramidal cells while all the layers contain stellate (nonpyramidal) cells. The cerebral cortex contains cytoarchitectonic fields that differ in the arrangement of cells and fibers. In different layers the pyramidal cells project to different parts of the nervous system. The cerebral cortex contains vertical columns perpendicular to the surface that subserve one sensory modality. In the somatosensory cortex this is exemplified by response to movement of a specific muscle, while in the visual cortex there are specific eye-dominance columns.

The brain and spinal cord are surrounded by three layers of meninges. The outermost layer is the dura mater, which is continuous with the periosteum of the flat bones of the skull. The next layer moving toward the neural tissue is the weblike arachnoid, which contains the main arteries and veins to the brain. The pia is the innermost layer and is invested by the glia limitans, which consists of astrocyte foot processes.

219. The answer is A. *(Burkitt, 3/e, p 129. Junqueira, 8/e, pp 171, 174, 177. Ross, 3/e, pp 286–287.)* The presence of large multipolar neurons with eccentrically placed nuclei and coarse granular Nissl bodies characterizes a neuron as autonomic rather than sensory. The autonomic nervous system mediates activity by two motor neurons placed in series: the first lies either in a nucleus of the brainstem or in the spinal gray matter; the second is located in a ganglion. The autonomic ganglia contain motor nerve cell bodies that convey impulses originating in the brain and spinal cord to smooth muscle and glands by way of the splanchnic nerves. Satellite cells surround neuronal cell bodies in ganglia and are similar to Schwann cells in function (e.g., insulation and metabolic regulation). The major difference between Schwann cells and satellite cells is that Schwann cells produce myelin in the peripheral nervous system.

220. The answer is D. *(Junqueira, 8/e, pp 175–179. Kandel, 3/e, p 264. Ross, 3/e, pp 283–284. Stevens, p 222.)* Axonal regeneration occurs in neurons if the perikarya survive following damage. The segment distal to the wound, including the myelin, is phagocytosed and removed by macrophages. The proximal segment is capable of regeneration since it remains in continuity with the perikaryon. Chromatolysis is the first step in the regeneration process in which there is breakdown of the Nissl substance, swelling of the perikaryon, and lateral migration of the nucleus of the neuron. Regeneration is dependent on the proliferation of Schwann cells, which serve to guide sprouting axons from the proximal segment toward the target organ that is being reinnervated. This process is referred to as *Wallerian regeneration.* Degeneration of perikarya and neuron processes occurs when there is extensive neuronal damage. Transneuronal degeneration only occurs when there is a single input (synapse) with another neuron. In the presence of multiple inputs, transneuronal degeneration does not occur.

221. The answer is E. *(Junqueira, 8/e, pp 164, 170. Kandel, 3/e, pp 19–20, 101. Stevens, pp 212–213.)* The nodes of Ranvier represent the space between adjacent units of myelination. This area is bare in the CNS, while in the PNS the axons in the nodes are partially covered by the cytoplasmic tongues of adjacent Schwann cells. Most of the Na^+ gated channels are located in the bare areas. Therefore, spread of depolarization from the nodal region along the axon occurs until it reaches the next node. This is often described as a series of jumps from node to node, or saltatory conduction. Increased efficiency is attained because energy-dependent Na^+ influx is limited to only the nodal regions.

222. The answer is C. *(Junqueira, 8/e, pp 160–162, 167–168. Kandel, 3/e, pp 1054–1057. Stevens, pp 213, 218.)* The blood-brain barrier is believed to be formed primarily from the presence of occluding junctions (zonulae occludentes) between endothelial cells that compose the lining of brain capillaries. The capillary endothelium is nonfenestrated, which also adds to the barrier. In addition, astrocytes form foot processes around the brain capillaries. Surrounding the CNS is a basement membrane with a lining of astrocyte foot processes; this forms the glia limitans, which also contributes to the integrity of the blood-brain barrier. Oligodendrocytes function in myelination of CNS axons. Microglia function as brain macrophages and are involved in antigen presentation and phagocytosis.

223. The answer is A. *(Alberts, 3/e, pp 540–541. Junqueira, 8/e, pp 186–189, 194. Kandel, 3/e, pp 43, 135–138. Stevens, pp 230–231.)* Neuromuscular, or myoneural, junctions represent the site at which end-feet (boutons terminaux) come in close proximity to the surface of muscle cells. The arrangement is similar to that found in a synapse and a neuromuscular junction can be considered the best-studied synapse. Na^+, K^+, and Cl^- voltage-gated channels are involved in the transmission of a nerve impulse but are not involved in the coupling of the action potential (an electrical signal) to neurotransmitter release (a chemical alteration). Ca^{2+} entry results in fusion of acetylcholine-containing synaptic vesicles with the presynaptic membrane and ultimately the release of neurotransmitter. Ca^{2+} influx into the end-feet may have a direct effect on phosphorylation of synapsin I, a vesicular membrane protein, which in its nonphosphorylated state blocks vesicle fusion with the presynaptic membrane.

224. The answer is D. *(Burkitt, 3/e, pp 372–373. Erlandsen, pp 61–62. Ross, 3/e, pp 296–297. Stevens, pp 223–224.)* Large, multipolar neurons in the cerebral cortex are pyramidal cells. Forming the stroma of the brain, the neuroglial cells, which are small, are located along the dendrites and soma of the neurons. Basket cells, Purkinje cells, and granule cells are located in the cerebellar cortex. The basket cells make profuse dendritic contact with the Purkinje cells, which have a very characteristic flask shape. The granule cells are small neurons located in the vicinity of the Purkinje cells.

225. The answer is A. *(Erlandsen, p 61. Junqueira, 8/e, pp 154, 156–157. Stevens, p 225.)* An axon is not evident in the histologic section accompanying the question. The axon arises from the axon hillock. Neither the axon nor axon hillock contains Nissl substance (rough endoplasmic reticulum), which is dispersed throughout the soma and dendrites. Dendrites generally are wider than axons, are of nonuniform diameter, and taper to a point. Motor neurons, such as the one illustrated, usually display large amounts of euchromatin and a distinct nucleolus characteristic of high synthetic activity.

226. The answer is D. *(Junqueira, 8/e, p 152. Kandel, 3/e, pp 902–904. Moore, Before We Are Born, 4/e, p 282. Moore, Developing Human, 5/e, pp 62–63, 385, 393.)* The neural crest consists of cells located lateral to the neural tube soon after its closure. These cells migrate throughout the developing embryo and make major contributions to adult neural and nonneuronal structures. The neural crest forms most of the peripheral nervous system in contrast to the neural tube, which is the embryonic source of the central nervous system. The sensory neurons of the cranial and spinal sensory ganglia (e.g., dorsal root ganglia), sympathetic chain ganglia, postganglionic sympathetic and parasympathetic fibers of the autonomic nervous system, cells of the pia and arachnoid, Schwann cells, and the satellite cells of the dorsal root ganglia are the neural elements derived from the neural crest. Nonneuronal structures formed from the neural crest include the melanocytes of the skin, odontoblasts of the teeth, derivatives of the branchial arch cartilages (e.g., the pinnae of the ear), and the adrenal medulla. The adrenal medulla represents postganglionic sympathetic fibers that respond to inputs from the preganglionic sympathetic fibers in the splanchnic nerves, and it therefore could be considered a neural structure. Ventral horn cells are derived from the neuroepithelium of the neural tube.

227. The answer is C. *(Junqueira, 8/e, p 167. Kandel, 3/e, pp 302–304, 1050–1054, 1057, 1059.)* Cerebrospinal fluid (CSF) is synthesized by the choroid plexuses at a rate of about 0.5 L/day. CSF is formed as an ultrafiltrate of the blood and is actively secreted into the subarachnoid space through the foramina of Luschka and Magendie. Absorption occurs through the arachnoid villi, which extend into the lumen of the subarachnoid space and form a passageway to the venous sinuses. Blockage of outflow from the median and lateral apertures or decreased absorption of CSF may result in hydrocephalus. A lumbar puncture is done between L2 and L3 in order to safely avoid the end of the spinal cord.

228. The answer is B. *(Burkitt, 3/e, p 129. Erlandsen, p 62. Junqueira, 8/e, pp 171, 174, 177. Ross, 3/e, pp 273, 286–287. Stevens, p 222.)* Ganglia are aggregations of nerve cells surrounded by dense fibrous connective tissue and contain loose connective tissue, perikarya, supporting cells (i.e., Schwann or satellite cells), and axons.

Ganglia are of two main types: sensory and autonomic. Sensory ganglia are associated with both spinal and cranial nerves and are pseudounipolar. Sympathetic ganglia are associated with multipolar neurons and include synapses between pre- and postganglionic fibers. There are no synapses in the sensory ganglia (e.g., dorsal root ganglia) since the pseudounipolar cells in these ganglia possess a T-shaped process with one process (dendrite) coming from receptors on the periphery and the other process extending into the central nervous system (e.g., spinal cord in the case of dorsal root ganglia).

229. The answer is A. *(Junqueira, 8/e, pp 160–162. Ross, 3/e, pp 267–268. Stevens, pp 213–214.)* Astrocytes are large, stellate cells that stain positively for the intermediate filament protein called *glial fibrillary acidic protein (GFAP)*. They form from the neuroepithelium and they have important functions in the developing and adult nervous systems. The microglia are the bone marrow–derived cells of the monocyte lineage found in the CNS. Astrocytes occur in two subtypes: protoplasmic and fibrous astrocytes. These subtypes differ in morphology and location. Fibrous astrocytes are found in the white matter while protoplasmic astrocytes are located in the gray matter of the CNS. The astrocytes may function as a potassium sink during prolonged neuronal activity; however, they have several better-defined functions. The astrocytes form a scaffolding for the migration of developing neurons during the differentiation of the embryonic nervous system. Astrocytes also form a scaffolding for neuronal elements in the adult brain. Many astrocytes in the adult CNS have extended foot processes called vascular end-feet which contact brain vascular elements. The function of the end-feet is probably associated with ion transport from the area surrounding the neurons to the blood vessels and maintenance of the blood-brain barrier. The CNS is surrounded by the glia limitans found beneath the pial surface; it is a basement membrane lined by astrocytes. After damage to the CNS, astrocyte proliferation is involved in scar formation (gliosis). Astrocytes are also the source of the most common glioma, astrocytoma.

230–233. The answers are 230-B, 231-A, 232-D, 233-C. *(Erlandsen, p 66. Junqueira, 8/e, pp 157–160. Kandel, 3/e, pp 130–133.)* There is no cytoplasmic continuity between adjacent neurons. Transmission from neuron to neuron occurs by either electrical or chemical synapses. Electrical synapses are gap junctions that allow ionic and current fluxes between neurons. In chemical synapses, neurotransmitter is released from the presynaptic membrane by exocytosis. The content of the vesicles crosses the synaptic cleft (between the pre- and postsynaptic membranes) and interacts with receptors on the postsynaptic membrane, which results in changes in the permeability of this membrane. Numerous mitochondria and synaptic vesicles are typically found on the presynaptic side of the synapse. The postsynaptic surface typically is more dense than the presynaptic membrane.

234–238. The answers are 234-A, 235-E, 236-C, 237-B, 238-B. *(Junqueira, 8/e, pp 160–164. Kandel, 3/e, pp 22–24, 39, 264. Ross, 3/e, pp 256–270. Stevens, pp 213– 216.)* The neuroglia, or glia, are the supportive cells of the nervous system. They are more numerous than neurons and unlike neurons they retain the capacity for cell proliferation under certain conditions. The glial cells may be subdivided into several types of cells: astrocytes, oligodendrocytes, ependymal and Schwann cells, and microglia. These cell types differ in embryologic origin, structure, and function.

Astrocytes stain positively for the intermediate filament protein *glial-fibrillary-acidic protein (GFAP).* They have essential functions in the developing and adult nervous system. Astrocytic proliferation is involved in scar formation during repair of damage to the brain, a process called *gliosis.* Other cells such as the microglia serve to remove debris, but the scar is formed by astrocytes. Microglia and astrocytes both make and respond to cytokines and growth factors during the repair process. Astrocytomas are the most common glioma and are derived from astrocytes.

Oligodendrocytes are derived from a separate cell lineage during development, but like the astrocytes they originate from the neuroepithelium of the neural tube. Oligodendrocytes function in the formation of the myelin in the CNS and can be identified immunocytochemically by the presence of myelin basic protein.

Microglia are the most controversial of the glial cells. They were originally identified in Golgi preparations of the CNS, but both their existence and origin have been questioned since their discovery. It is now well documented that the microglia not only exist but play an important role as brain macrophages. These are small cells with elongate nuclei and a thorny appearance. They may be recognized by antisera to common leukocyte antigen and MHC class II histocompatibility antigens, which document their origin from the bone marrow (specifically from monocytes) and their role in antigen presentation. When damage occurs in the brain, microglia are responsible for phagocytosis of debris. In cases of substantial damage to the blood-brain barrier (e.g., a stab wound), new monocyte-derived cells may enter the brain to function in phagocytosis and cytokine production. The cytokines include growth factors such as transforming growth factor-beta, which modulates synthesis of extracellular matrix, and chemokines, which serve as chemoattractants of monocytes. These factors are key elements in the regulation of events involved in wound healing.

Schwann cells are responsible for myelination in the peripheral nervous system. They are derived from the neural crest and are close relatives to the satellite cells of the dorsal root and autonomic ganglia. These cells surround the length of all axons in the peripheral nervous system and are sometimes known as neurilemmal, or sheath, cells.

Ependymal cells are the cells remaining in the ventricular layer of the developing neural tube after neurons and glia have differentiated. These cells form an epithelium that may possess both microvilli and cilia. There are junctional complexes with desmosomes but no occluding junctions between most ependymal cells except in the choroid plexuses. The absence of tight junctions allows communication of the cerebrospinal fluid with the brain parenchyma.

239–243. The answers are 239-F, 240-C, 241-G, 242-A, 243-E. *(Junqueira, 8/e, pp 164–167. Ross, 3/e, pp 281–282. Stevens, pp 216–217.)* The meninges are the protective layers of the brain and spinal cord. The periosteum is the connective tissue surrounding the bone of the skull. This layer retains osteogenic potential even in the adult. The dura mater ("tough mother") is composed of dense connective tissue and possesses very limited osteogenic potential. The dura is the site of the venous sinuses. In the spinal cord, the dura is separated from the periosteum by the epidural space. The thin subdural space lies between the dura mater and the arachnoid. The arachnoid is composed of a weblike avascular connective tissue, which forms villi for the reabsorption of cerebrospinal fluid (CSF) into the venous sinuses found in the dura. The subarachnoid space contains the CSF, which is formed both by ultrafiltration of the blood and transport across the epithelial lining of the choroid plexuses. The pia covers the brain and spinal cord as a delicate, vascular connective tissue. It lines the perivascular spaces through which blood vessels penetrate the CNS.

Cardiovascular System, Blood, and Bone Marrow

DIRECTIONS: Each question below contains five suggested responses. Select the **one best** response to each question.

244. Vasa vasorum provide a function analogous to that of

(A) valves
(B) basal lamina
(C) coronary arteries
(D) endothelial diaphragms
(E) arterioles

245. Which of the following types of collagen are found in the blood vessels?

(A) Types I, III and IV
(B) Types I and II
(C) Type I only
(D) Types I and IV
(E) Types II and IV

246. The blood cell that is a precursor to microglia, Kupffer cells, and Langerhans cells is

(A) a basophil
(B) an eosinophil
(C) a neutrophil
(D) a monocyte
(E) a lymphocyte

247. Glanzmann's disease (also known as *thrombasthenia*) is a disease in which there is a defect in the glycoprotein IIb-IIIa complex, resulting in a defect in β_3-integrin structure on platelets. In these patients, platelets

(A) are reduced in size
(B) have limited ability to secrete platelet-derived growth factor (PDGF)
(C) are unable to undergo normal shape changes
(D) demonstrate poor aggregation
(E) demonstrate accelerated hemostasis

Questions 248–252

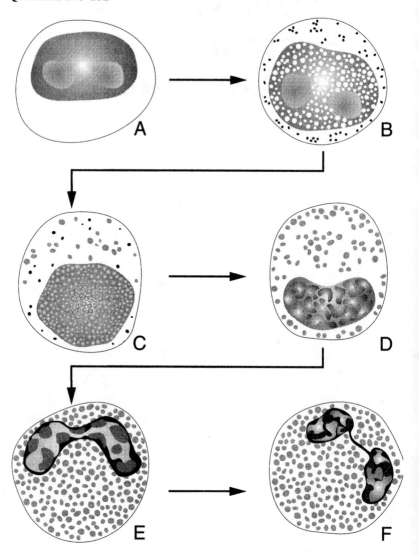

248. The cell labeled A is best described as a

(A) myeloblast
(B) proerythroblast
(C) metamyelocyte
(D) myelocyte
(E) promyelocyte

249. The mature blood cell labeled F functions primarily during

(A) allergic and parasitic infections
(B) immediate hypersensitivity reactions
(C) antibody production
(D) graft rejection
(E) phagocytosis of bacteria

250. During embryologic development, the earliest site in which the process shown in the figure occurs is the

(A) bone marrow
(B) yolk sac
(C) liver
(D) spleen
(E) kidney

251. In the adult, the best place to sample for cells of this lineage would be the

(A) sternum
(B) fibula
(C) humerus
(D) tibia
(E) scapula

252. The granules that develop during differentiation of this cell type contain

(A) heparin
(B) histamine
(C) histaminase
(D) platelet-derived growth factor (PDGF)
(E) thymosin

253. Maintenance of diastolic pressure and adaptation to systolic pressure within the circulatory system are related to the presence of

(A) smooth muscle in arterioles
(B) elastic fibers in large arteries
(C) type I collagen in the adventitia
(D) valves in large veins
(E) thicker walls in the pulmonary than the systemic arteries

254. Which of the following structures best represents the large endothelial pores defined in physiologic studies?

(A) Tight junctions
(B) Gap junctions
(C) Pinocytotic vesicles (fenestrae)
(D) Zonula adherens
(E) Desmosomal junctions

255. In fenestrated capillaries with diaphragms,

(A) fenestrations allow for free passage of small cells
(B) diaphragms restrict the passage of anionic molecules
(C) transport is independent of vesicular mechanisms
(D) transport is independent of channel mechanisms
(E) transport of macromolecules is dependent on a paracellular pathway

256. Anionic proteins such as insulin and transferrin cross capillary endothelial cells primarily via

(A) diffusion through the membrane
(B) intercellular space
(C) fenestrated diaphragms
(D) gap junctions
(E) vesicular and channel pathways

257. In diabetes mellitus, thickening of the endothelial basal lamina is most likely due to

(A) decreased proliferation of endothelial cells
(B) increased synthesis of type III collagen
(C) increased production of elastic fibers
(D) decreased collagen degradation due to altered glycosylation
(E) decreased cross-linking of collagen

258. Distribution of blood to specific organs is regulated predominantly by

(A) muscular arteries
(B) conducting arteries
(C) the capillary bed
(D) postcapillary venules
(E) medium-sized veins

259. Blood flow is slowest in the

(A) aorta
(B) muscular arteries
(C) veins
(D) capillaries
(E) arterioles

260. Weibel-Palade granules contain von Willebrand factor (factor VIII). Which of the following statements is true in regard to these granules and their content?

(A) They are only found in the endothelium of capillaries

(B) Absence of these granules leads to enhanced platelet aggregation in endothelial cells

(C) Deficiency may lead to hemophilia

(D) Von Willebrand factor is primarily responsible for the conversion of proendothelin to endothelin

(E) Von Willebrand factor is a monoamine oxidase that inactivates norepinephrine and serotonin

261. Which of the following statements about pericytes is true?

(A) They are closely associated with fibroblasts

(B) They form the tunica intima of capillaries

(C) They do not have contractile properties

(D) They are located primarily in conducting arteries and large veins

(E) They have the potential for mitogenic activity as well as transformation into other cell types

262. Factors that contribute to the antithrombogenic activities of endothelial cell surfaces include

(A) production of prostacyclin

(B) positive electrostatic charge on the endothelial cell surface

(C) production of thromboxane by platelets

(D) allowing platelets to move into the vascular wall

(E) synthesis of plasminogen inhibitor

263. Organs such as the brain and thymus have a more effective blood barrier because their blood capillaries are of the

(A) continuous type with only a few endothelial plasmalemmal vesicles

(B) fenestrated type with diaphragms

(C) fenestrated type without diaphragms

(D) discontinuous type with diaphragms

(E) discontinuous type without diaphragms

264. Atherosclerosis is usually initiated by

(A) proliferation of smooth muscle cells

(B) formation of an intimal plaque

(C) attraction of platelets to collagen microfibrils

(D) adventitial proliferation

(E) injury to the endothelium

265. The blood vessel in the accompanying electron micrograph is

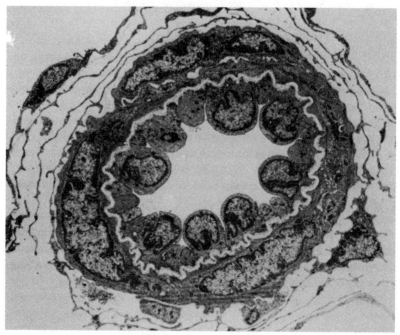

Courtesy of Dr. Vincent H. Gattone II.

(A) an arteriole
(B) a venule
(C) a medium artery
(D) an elastic artery
(E) a capillary

266. In the accompanying photomicrograph, the several larger cells (containing eccentric nuclei, radially arranged coarse chromatin, and basophilic cytoplasm) normally function in

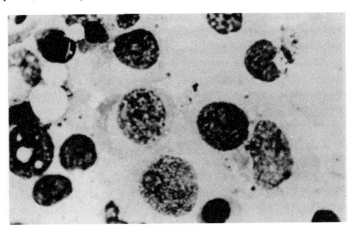

(A) antigen recognition
(B) cellular immunity
(C) histamine release
(D) immunoglobulin synthesis
(E) phagocytosis

267. If fresh human blood is centrifuged in the presence of an anticoagulant such as heparin to obtain a hematocrit, the resulting fractions are

(A) serum, packed erythrocytes, and leukocytes
(B) leukocytes, erythrocytes, and serum proteins
(C) plasma, buffy coat, and packed erythrocytes
(D) fibrinogen, platelets, buffy coat, and erythrocytes
(E) albumin, plasma lipoproteins, and erythrocytes

268. Thrombocytopenia is a reduction in the number of circulating blood platelets. Which of the following would most likely occur in thrombocytopenia?

(A) Decreased vascular permeability
(B) Failure of initiation of the blood clotting cascade
(C) Failure of conversion of fibrinogen to fibrin
(D) Absence of plasmin
(E) Absence of thrombin

269. Erythrocytes may have abnormal shapes and sizes in certain diseases. In iron deficiency you would expect to see

(A) microcytic, hypochromatic anemia with smaller mature erythrocytes
(B) macrocytic, hyperchromatic anemia with fewer, larger mature erythrocytes
(C) poikilocytosis (shape change) and more fragile erythrocytes
(D) spherical rather than biconcave erythrocytes
(E) no change in erythrocyte size or shape, but a substantial drop in the hematocrit

270. The hematopoietic growth factor that is synthesized in the kidney and stimulates the formation and release of reticulocytes into the blood is

(A) hemonectin
(B) erythropoietin
(C) interleukin 3 (IL-3)
(D) granulocyte-macrophage colony stimulating factor (GM-CSF)
(E) macrophage colony stimulating factor (M-CSF)

271. In a bone marrow transplant, the seeding of intravenously injected hematopoietic precursors is specifically and directly regulated by

(A) a lectin-glycoconjugate interaction with galactosyl specificity
(B) hematopoietic stromal cell interactions
(C) stromal cell proliferation
(D) endothelial cell engulfment of hematopoietic precursors
(E) cell surface–proteoglycan interactions

DIRECTIONS: Each numbered question or incomplete statement below is NEGATIVELY phrased. Select the **one best** lettered response.

272. After loss or damage to the endothelium of a blood vessel, platelets carry out all the following functions EXCEPT

(A) inhibition of endothelial, smooth muscle, and fibroblast cell proliferation

(B) molding of their shape and adherence to the surface through the action of the cytoskeleton

(C) secretion of coagulation factors and serotonin

(D) clot retraction

(E) clot dissolution

273. All the following statements are true regarding transcytosis EXCEPT

(A) it involves pinocytosis

(B) it occurs by nonspecific absorption

(C) it involves receptor-mediated processes

(D) it involves the entry of molecules into both endosomal and lysosomal compartments

(E) it is an active process

274. Metabolic functions of endothelial cells include all the following EXCEPT

(A) conversion of angiotensin I to angiotensin II

(B) inactivation of bradykinin

(C) production of type III collagen

(D) antithrombogenic activity

(E) production of coagulation and anticoagulation factors

275. Contents of the plasma include all the following EXCEPT

(A) albumin

(B) fibrinogen

(C) immunoglobulins

(D) alpha and beta globulins

(E) platelets

276. In the human fetus, hematopoiesis takes place in all the following during the third and fourth months of gestation EXCEPT the

(A) yolk sac

(B) liver

(C) spleen

(D) thymus

(E) bone marrow

DIRECTIONS: Each group of questions below consists of lettered headings followed by a set of numbered items. For each numbered item select the **one** lettered heading with which it is **most** closely associated. Each lettered heading may be used **once, more than once, or not at all.**

Questions 277–282

Match each description with the correct cell type.

(A) Plasma cells
(B) Eosinophils
(C) Basophils
(D) Neutrophils
(E) Monocytes
(F) B lymphocytes
(G) T lymphocytes

277. Cells that increase in number in allergic reactions and parasitic infections

278. Phagocytosis of bacteria

279. Secretion of histaminases that inactivate histamine

280. Major component of pus

281. Immediate hypersensitivity reactions during exposure to allergen through release of its granules

282. Structure and granular content similar to that of mast cells

Questions 283–287

Match each blood cell with its labeled representation in the figure below.

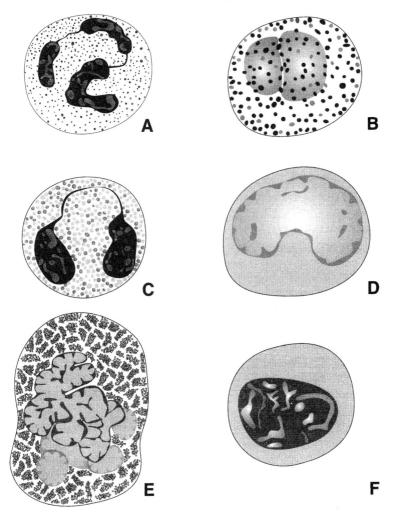

283. Neutrophil

284. Megakaryocyte

285. Monocyte

286. Lymphocyte

287. Basophil

Cardiovascular System, Blood, and Bone Marrow
Answers

244. The answer is C. *(Junqueira, 8/e, pp 202–204. Ross, 3/e, pp 306–308. Stevens, pp 111, 114.)* Vasa vasorum are vessels within a vessel and are found primarily in the adventitia of large arteries and veins. They serve to provide nutrition and oxygenated blood to the thick media and adventitia of these vessels, which are unable to obtain nutrition by diffusion from the lumen. Coronary arteries fulfill a similar function for the myocardium.

245. The answer is A. *(Junqueira, 8/e, p 202. Ross, 3/e, p 100. Stevens, p 45.)* Type I collagen is found in the adventitia of the blood vessels. Smooth muscle cells in the media of blood vessels synthesize primarily type III collagen as well as fibronectin, glycosaminoglycans, and elastin. Type IV collagen is associated with basement membranes (e.g., underlying the endothelium). Type II collagen is found primarily in hyaline and elastic cartilage.

246. The answer is D. *(Junqueira, 8/e, pp 103–107, 163, 230–232, 244, 250–251, 310, 351. Ross, 3/e, pp 200–201, 339. Stevens, pp 76, 84–85, 354.)* The system once called the *reticuloendothelial system* is now known as the *monocyte-macrophage,* or *mononuclear phagocyte, system.* Members of this system are derived from the monocyte lineage at some time during development. Microglia are the macrophages of the brain, Kupffer cells are the macrophages of the liver, and Langerhans cells are the macrophages of the skin. These cells are responsible for phagocytosis and antigen presentation. They also secrete and respond to the intercellular mediators known as *cytokines.*

247. The answer is D. *(Alberts, 3/e, p 996. Isselbacher, 13/e, pp 317, 321, 1802.)* Glanzmann's disease, or thrombasthenia, is caused by a defect in the β_3-integrins. The result is the loss of normal binding to fibrinogen and loss of normal aggregation of platelets. The effect on the patient is delayed clotting time, reduced hemostasis, and therefore excessive bleeding following injury. *Hemostasis* is the term used for the body's response to endothelial disruption leading to the entrance of blood into the subendothelial connective tissue. It is often divided into two phases: (1) primary, which is the formation of a platelet plug; and (2) secondary, which results in the formation of the clot, composed of fibrin. Granule release, size, and normal shape changes do not appear to be affected in Glanzmann's disease.

248–252. The answers are 248-A, 249-A, 250-B, 251-A, 252-C. *(Junqueira, 8/e, pp 219, 222, 224–228, 235, 241–242, 246. Ross, 3/e, pp 197–198, 202, 205. Stevens, pp 67, 69–70, 72.)* The lineage shown in the figure illustrates eosinophilic development. The first stage in granulopoiesis is the myeloblast (A), a large cell with prominent light-staining nucleoli with only a little cytoplasm, generally without granules. The promyelocyte (B) is the next cell in the lineage. It is larger than the myeloblast, nucleoli are less visible, and primary granules are present in the cytoplasm. In the next stage, myelocyte-specific granules (i.e., eosinophilic, neutrophilic, or basophilic) are seen in the cytoplasm with flattening of the nucleus. The eosinophilic myelocyte (C) differentiates into the eosinophilic metamyelocyte (D) when invagination of the nucleus begins. Further invagination leads to the formation of an eosinophilic band (E) and ultimately a mature eosinophil (F). An eosinophil has a bilobed nucleus and plays an important role in allergic and parasitic infections. The granules stain with eosinophilic dyes and contain major basic protein, histaminase, peroxidase, and some hydrolytic enzymes. This cell has an affinity for antigen-antibody complexes and, although phagocytic, it is not as active against bacteria as neutrophils. Eosinophils secrete histaminase and counteract the release of histamine from basophils and mast cells. Mast cells and basophils are essential in hypersensitivity reactions; B lymphocytes differentiate into antibody-producing plasma cells; T lymphocytes are primarily responsible for graft rejection; and neutrophils are responsible for phagocytosis of bacteria.

Hematopoiesis occurs in the flat bones of the skull and other bones such as the sternum and iliac crest of the adult human. It also occurs in the long bones during development, but many of these areas become dominated by yellow marrow that contains many fat cells (adipose tissue). However, during development the yolk sac is the first site of hematopoiesis. Later in development, blood cell formation occurs in the spleen and liver before hematopoiesis becomes limited to the bone marrow. The kidney produces erythropoietin, a growth factor that stimulates red cell development, but it is not considered a site of blood cell development. Although most bones in the body are involved in hematopoiesis during growth, the marrow of the sternum, ribs, vertebrae, skull, and proximal femora are the primary sites of blood cell development by the time that skeletal maturity is achieved. The inactive yellow marrow found in most long bones can be reactivated upon exposure to the proper stimulus (i.e., severe blood loss).

253. The answer is B. *(Junqueira, 8/e, pp 209–211. Ross, 3/e, p 303. Stevens, p 113.)* The conducting arteries are required to adjust to the force of ventricular contraction during systole (120 to 160 mmHg) followed by sudden relaxation (60 to 90 mmHg) during diastole. The presence of extensive elastic fibers permits rapid arterial wall stretch and relaxation and maintenance of arterial blood pressure. Blood is ejected from the left ventricle into the large arteries only during systole; however, blood flow is uniform because of the elasticity of the large, conducting arteries. Pulmonary arteries have thinner walls than the systemic arteries.

254. The answer is C. *(Junqueira, 8/e, pp 204–209.)* Endothelial cells are joined together by tight junctions (zonula occludens) with a rare desmosome (macula adherens) observed between the cells. Gap junctions may be present between adjacent endothelial cells and permit transfer of information between adjoining cells. Intercellular junctions, particularly the tight junctions, function as the small endothelial pores (approximately 10 nm in diameter) observed in physiologic studies. The large pores (50 to 70 nm in diameter) are represented by the pinocytotic vesicles (fenestrae).

255. The answer is B. *(Junqueira, 8/e, pp 205–209. Simionescu, Ann N Y Acad Sci 401:9–24, 1982.)* Capillary endothelia are continuous, fenestrated, or discontinuous (sinusoids). Transcellular openings known as *fenestrae* occur in many of the visceral capillaries. In other cases, such as the hematopoietic organs, there are large gaps in the endothelium and the capillaries are classified as discontinuous. Diaphragms contain proteoglycans, particularly high concentrations of heparan sulfate. This results in numerous anionic sites that repel anionic proteins. The diaphragms do facilitate the passage of water and small molecules dissolved in fluid. Fenestrations do not permit the passage of cells. Other transport mechanisms occur in fenestrated endothelia and represent different aspects of the dynamic movement of molecules across endothelia, including vesicular and channel mechanisms. Except under pathologic conditions, intercellular junctions do not allow the passage of large proteins. The paracellular pathway through the intercellular junctions is a mechanism for passage of water and small dissolved molecules.

256. The answer is E. *(Simionescu, Ann N Y Acad Sci 401:9–24, 1982.)* Plasmalemmal vesicles and channel diaphragms are neutrally charged and rich in galactose and N-acetylglucosamine. Vesicular and channel pathways are required for transport of anionic proteins such as insulin, transferrin, albumin, and low-density lipoprotein (LDL).

257. The answer is D. *(Alberts, 3/e, pp 980–981. Cotran, 5/e, pp 915–921. Isselbacher, 13/e, p 1996.)* In diabetes mellitus, increased blood glucose levels lead to altered glycosylation. During collagen synthesis, glycosylation is essential for the appropriate cross-linking of collagen. Collagenases responsible for the specific degradation of collagens are also dependent on correct cross-linking. The opposite situation is observed in cases of scurvy, in which the absence of vitamin C leads to inhibition of proline hydroxylation, abnormal procollagen helix formation, and rapid degradation of defective pro-α-chains. Cross-linking of collagen in blood vessels increases with age. Decreased cross-linking of collagen would probably facilitate degradation.

258. The answer is A. *(Junqueira, 8/e, pp 209–211.)* The muscular arteries are also known as *distributing arteries* and regulate blood flow to organs. The state of contraction of the muscular arteries is regulated by local factors as well as innervation by sympathetic fibers. The degree of contraction regulates blood flow distribution between organs. When the tunica media of the muscular artery is contracted, less blood flow occurs to the organ. In a more relaxed state, there is increased blood flow to the same organ.

259. The answer is D. *(Junqueira, 8/e, pp 209–211.)* The variations between systole and diastole are balanced by the elasticity of the large arteries. Therefore, arterial blood pressure and the speed of blood flow decrease as one moves further from the heart. This coincides with a decrease in the elasticity of the vessels as one moves from larger to smaller arteries. Also, as the blood vessel diameter decreases, the velocity of blood flow is reduced.

260. The answer is C. *(Cotran, 5/e, pp 58, 100–101, 469. Junqueira, 8/e, p 209.)* Weibel-Palade granules contain von Willebrand factor (factor VIII). These granules are located in the endothelium of vessels larger than capillaries. A deficiency of factor VIII leads to decreased platelet aggregation and hemophilia.

261. The answer is E. *(Junqueira, 8/e, pp 205, 217.)* Pericytes are found in capillaries and postcapillary venules and represent the tunica media of these vessels. They surround endothelial cells and contain large amounts of actin, myosin, and tropomyosin as detected by immunohistochemical methods. Following injury, pericytes revert to a mesenchymal type of stem cell and are capable of extensive cell proliferation or differentiation into other cell types or both. The transforming ability of pericytes is observed in vascular tumors called *hemangiopericytomas*.

262. The answer is A. *(Junqueira, 8/e, pp 208–209, 233. Stevens, p 116.)* Thrombi are often formed after damage to the endothelium. The absence of the antithrombogenic endothelial cells results in aggregation of platelets followed by fibrin formation. The antithrombogenic properties of the endothelium include production of prostacyclin, which functions through a cyclic AMP second messenger to inhibit thromboxane production by platelets. The synthesis of plasminogen activator also contributes to the antithrombogenic activity of the endothelium. Von Willebrand factor (factor VIII) stimulates blood clotting.

263. The answer is A. *(Burkitt, 3/e, pp 134, 201. Junqueira, 8/e, pp 161, 167, 205–206, 208, 253–254. Ross, 3/e, pp 256, 347–348. Stevens, pp 89, 218.)* The blood-thymus barrier provides the appropriate microenvironment for education of T cells without exposure to self. The capillary endothelium is continuous, as is the basal lamina. The capillary is further surrounded by perivascular connective tissue and epithelial cells and their basement membrane. In the case of the blood-brain barrier, there is also a continuous endothelium with an absence of fenestrations and basal lamina. Surrounding the basal lamina are the foot processes of astrocytes, which form the glia limitans. In addition to this aspect of the blood-brain barrier, endothelial cells are joined by tight occluding junctions with many sealing strands.

264. The answer is E. *(Junqueira, 8/e, pp 211, 233. Stevens, p 116.)* Atherosclerosis is one form of arteriosclerosis (hardening of the arteries) that involves deposition of fatty material in the walls of primarily the conducting arteries. The process is initiated through damage to the endothelial cells, which exposes the subjacent connective tissue (subendothelium). The loss of the antithrombogenic endothelium results in aggregation of platelets. The intima and media become infiltrated with lipid. Intimal thickening occurs through the addition of collagen and elastin with an abnormal pattern of elastin crosslinking. Platelets release mitogenic substances that stimulate proliferation of smooth muscle cells. The thickening of the intima is also called an *atheromatous plaque* and worsens with repeated damage to the endothelium. It is most dangerous in small vessels, particularly the coronary arteries, where occlusion can result in a myocardial infarction. Atherosclerotic plaques also lead to thrombi and aneurysms.

265. The answer is A. *(Burkitt, 3/e, pp 143–145. Erlandsen, pp 71–72. Junqueira, 8/e, pp 204–205, 209–211, 213.)* The blood vessel in the electron micrograph is a small artery, or arteriole. There is only one layer of smooth muscle, but a distinct internal elastic membrane is present. Muscular (medium) arteries contain more smooth muscle and elastic (large) arteries contain extensive elastic tissue. There is no visible internal elastic membrane in a venule. A capillary lacks smooth muscle and is composed only of a single layer of endothelial cells.

266. The answer is D. *(Alberts, 3/e, pp 1198–1199. Burkitt, 3/e, pp 74, 192, 196. Erlandsen, pp 21, 28, 77. Junqueira, 8/e, pp 107–111.)* The cells illustrated in the photomicrograph accompanying the question are plasma cells. These cells are the end stage of differentiation of B lymphocytes. Plasma cells are characterized by eccentric nuclei with coarse granules of heterochromatin arranged in a radial pattern about the nuclear envelope. Membrane-bound ribosomes are extremely plentiful, providing the cytoplasm with a characteristic intense basophilia. This rough endoplasmic reticulum is involved in antibody production, principally immunoglobulin G (IgG). The juxtanuclear region, which does not stain, represents the Golgi complex, in which the antibodies are processed for secretion.

267. The answer is C. *(Junqueira, 8/e, p 218.)* When blood is removed from the body, it forms a clot that contains platelets, erythrocytes, leukocytes, and a clear, yellow fluid known as *serum.* Hematocrit is the volume of erythrocytes per unit volume of blood (e.g., 40 to 50 percent in adult human males). When centrifuged with anticoagulants, blood separates into three layers: plasma, buffy coat (a thin white layer consisting of leukocytes found immediately above the lowest layer), and the packed erythrocyte layer at the bottom of the tube.

268. The answer is C. *(Junqueira, 8/e, pp 232–233. Stevens, pp 78–79.)* Platelets (thrombocytes) are cell fragments obtained by the breakup of megakaryocytes. These cell fragments contain a number of important substances as well as cytoskeletal elements involved in biologic processes such as clot retraction. Platelets function in aggregation, coagulation, clot retraction, and removal. They are involved in the conversion of fibrinogen to fibrin through the action of platelet phospholipids. Thrombin is also involved in this conversion, but it is a plasma protein, not a platelet secretory factor. Platelets are not required for the initiation of the blood clotting cascade, but they are required for the adherence and normal formation of a clot. Plasmin is not secreted by platelets but is formed by the conversion of plasma-derived plasminogen under the influence of plasminogen activator secreted by endothelial cells. Plasmin is involved in dissolution, not formation, of blood clots. Thrombocytopenia is a reduction in the number of platelets. Under this condition, fibrinogen will not be converted to fibrin in sufficient quantity to allow normal clotting. The absence of platelet aggregation interferes with the normal maintenance and repair of endothelial injury. The endothelium becomes increasingly leaky and eventually may permit thrombocytopenia purpura with seepage of blood from the vessel.

269. The answer is A. *(Junqueira, 8/e, p 221. Stevens, pp 69, 71.)* In iron deficiency, anemia results with the presence of smaller, pale-staining erythrocytes (microcytic, hypochromatic). In hemolytic anemia there is excessive destruction of red blood cells in the spleen. Hyperchromic, macrocytic anemia results from vitamin B_{12} deficiency. The presence of spherical rather than biconcave erythrocytes is associated with spherocytosis, which often results in hemolysis. The membrane undergoes deformation due to the inability of ankyrin to bind spectrin.

270. The answer is B. *(Isselbacher, 13/e, pp 1793–1798. Junqueira, 8/e, pp 234–237. Tavassoli, Proc Soc Exp Biol Med 196:367–373, 1991.)* Cells in the bone marrow are derived from a single type of stem cell. As the stem cell proliferates, its potentiality is reduced, so that clones derived from the pluripotential stem cell become gradually more specialized in terms of the differentiated cells that can originate from that clone. These cell groupings are called colonies and are described as colony-forming units (CFUs). CFUs include erythrocytic, lymphocytic, monocytic, granulocytic, and basophil-derived colonies.

Hematopoietic growth factors include the granulocytic growth factors: granulocyte colony stimulating factor (G-CSF), granulocyte-macrophage colony stimulating factor (GM-CSF), and macrophage colony stimulating factor (M-CSF), all of which stimulate the synthesis of cells derived from monocyte colony forming units (M-CFUs) and monocyte-granulocyte colony forming units (MG-CFUs). Erythropoietin is synthesized by the peritubular cells of the kidney cortex and stimulates the differentiation of cells from the erythrocyte colony forming units (E-CFUs). This growth factor is particularly effective in stimulating the later stages of erythropoiesis (e.g., the differentiation and release of reticulocytes from the bone marrow). The earlier steps include differentiation of proerythroblasts, basophilic erythroblasts before the formation of reticulocytes and finally erythrocytes.

Differentation of myeloid cells, both granulocytes and monocytes, is stimulated by GM-CSF, M-CSF, G-CSF, and IL-3. Colony stimulating factors are named for their activities. GM-CSF and IL-3 are general in their stimulation of white cell precursors; they stimulate both granulocytes and macrophages. M-CSF and G-CSF selectively stimulate macrophage and granulocyte differentiation, respectively. In the granulocytic series, differentiation follows these steps: myeloblast, promyelocyte, myelocyte, metamyelocyte, and mature granulocyte cell lineage.

The hematopoietic growth factors are produced by endothelial or stromal cells of the bone marrow. In addition some of these growth factors are produced by macrophages and lymphocytes. G-CSF and GM-CSF also activate mature cells and increase their sensitivity to chemotactic factors. These factors are becoming useful in the treatment of patients following anticancer radiation or chemotherapeutic treatments to reverse leukocytopenia or deficiencies of specific cells derived from the bone marrow.

Extracellular matrix proteins such as laminin, fibronectin, and hemonectin (a bone marrow–specific cell adhesion molecule) also play a role in differentiation of colony forming cells in the bone marrow. The release of the cells from the bone marrow is regulated by these growth factors and other hormones that interfere with the interaction between integrins on the cell surface and extracellular matrix molecules. Glucocorticoids and some complement components stimulate release of bone marrow cells when they reach an appropriate stage.

271. The answer is A. *(Alberts, 3/e, pp 1164–1173. Cotran, 5/e, pp 194–195. Junqueira, 8/e, pp 234–237. Tavassoli, Proc Soc Exp Biol Med 196:367–373, 1991.)* When a bone marrow transplant is undertaken, the hematopoietic cells seed, or home, to the bone marrow where they proliferate and differentiate into granulocytes, macrophages, and some erythroid cells. This process requires the presence of a lectin on the surface of the marrow sinus endothelial cells that recognizes marrow progenitor cells and initiates the transport of these cells across the endothelium. The lectin-glycoconjugate interaction is galactosyl-specific.

The presence of a homing receptor on the surface of a hematopoietic precursor is required for interaction with the stromal cells of the bone marrow. The presence of the homing receptor on the cell surface acts as a lectin and binds both galactose and mannose residues simultaneously on the surface of the bone marrow stromal cells. In the absence of this interaction, precursors differentiate into erythroid precursors. The homing receptors and the interactions with the stromal cells are therefore required for differentiation into macrophage and granulocyte progenitors.

There are various factors in the extracellular matrix that influence the proliferation and differentiation of hematopoietic stem cells: fibronectin, hemonectin, proteoglycans, and colony stimulating factors. These factors are not required for the lectin-glycoconjugate interaction that must occur for proper seeding of the cells injected in a bone marrow transplant.

272. The answer is A. *(Guyton, 8/e, pp 390–394. Junqueira, 8/e, pp 233, 245–246. Ross, 3/e, pp 190–192. Stevens, pp 78–79.)* Platelets are derived from megakaryocytes and lack nuclei. They contain specific cytoplasmic elements for storage of various secretory products: serotonin, coagulation factors, von Willebrand's factor, thromboxane, and ADP. When the endothelium is injured, platelets can react with collagen and damaged endothelial cells to form a plug. The cytoskeleton of the platelet contains actin and myosin as well as thrombosthenin. This extensive cytoskeleton assists in changes in shape of the platelet as well as contractions, which assist in the release of secretory granules. Some of these products, such as ADP and thromboxane, are produced by the platelets; others are absorbed from the plasma (serotonin). The secretory products are involved in the clotting cascade as well as the aggregation and attraction of other platelets. Platelet-derived growth factor is released by platelets and stimulates the proliferation of endothelial cells, vascular smooth muscle cells, and fibroblasts. Secretory products are released through an open canalicular system, which reaches invaginations in the plasma membrane. Platelets are also involved in clot dissolution through the action of acid hydrolases and clot retraction.

273. The answer is D. *(Alberts, 3/e, pp 623–625. Junqueira, 8/e, pp 204–208. Ross, 3/e, pp 37, 465.)* Transport through endothelial cells occurs by a number of mechanisms: cell membrane pathway, vesicular pathway, intercellular junctions (paracellular pathway), channel pathway, and open fenestrae (e.g., glomerular capillaries) or closed fenestrae (e.g., visceral capillaries such as those of the gastrointestinal tract). Transcytosis is an active process for transport of molecules from the extracellular space on the luminal side of the cell to the extracellular space on the abluminal side of the cell. The transport across the cell bypasses endosomes and lysosomal systems. Endothelial vesicular uptake may make use of pinocytosis (e.g., albumen and glycogen), nonspecific absorption (e.g., heme), or receptor-mediated transcytosis (e.g., LDL).

274. The answer is C. *(Cotran, 5/e, pp 469–470. Isselbacher, 13/e, pp 946–947. Junqueira, 8/e, pp 202, 208–209. Simionescu, Ann N Y Acad Sci 401:9–24, 1982.)* Endothelial cells subserve a number of important functions. They synthesize the basal lamina including types IV, V, and VIII collagens, fibronectin, and laminin. Type III collagen is found in reticular fibers synthesized by smooth muscle cells of the blood vessel wall. Secretion of A and B blood group antigens also occurs in endothelial cells. Angiotensin converting enzyme on the endothelial cell surface converts angiotensin I to angiotensin II (a potent vasoconstrictor) but also serves as an inactivation enzyme (bradykininase) for bradykinin, a vasodilator. The endothelium produces nitric oxide, also known as *endothelium-derived relaxing factor (EDRF),* and endothelin, the most potent vasoconstrictor in the body. Endothelial cells produce plasminogen activator and prostacyclin, which are potent anticoagulants, and plasminogen inhibitor, a coagulant.

275. The answer is E. *(Junqueira, 8/e, p 219. Ross, 3/e, pp 188–189.)* Plasma proteins include albumin; alpha, beta, and gamma globulins (immunoglobulins); and fibrinogen. In addition 10 complement proteins and kinins are components of the plasma. Platelets are formed elements in the blood and are not part of the plasma. They are found in the upper part of the buffy coat layer when one is obtaining a hematocrit.

276. The answer is A. *(Junqueira, 8/e, p 234.)* During prenatal development the first site of blood cell development (hematopoiesis) is extraembryonic—in the yolk sac. This site is replaced by the liver (during the second month) and subsequently by the bone marrow, which begins to function in the second month and becomes the predominant hematopoietic site during months 5 to 9 of gestation. The thymus is responsible for the education of T cells after the second month of gestation, while the spleen is involved specifically in the production of red blood cells (erythropoiesis) from months 2 to 5 of gestation.

277–282. The answers are 277-B, 278-D, 279-B, 280-D, 281-C, 282-C. *(Junqueira, 8/e, pp 103, 107–109, 222–232. Ross, 3/e, pp 192–200, 207. Stevens, pp 71–77.)* The four major components of blood are red blood cells (erythrocytes), white blood cells (leukocytes), platelets, and plasma. The white blood cells are subdivided into five cell types: lymphocytes, monocytes, and the three types of granulocytes—neutrophils, eosinophils, and basophils. Generally neutrophils, which are also called *polymorphonuclear leukocytes,* are the most common white blood cell found in the bloodstream and basophils are the least numerous.

Neutrophils are involved in the acute phase of inflammation and are responsible for the phagocytosis of invading bacteria. They are attracted to bacteria and must bind to bacteria before phagocytosis can occur. Chemoattractants include complement fragments and *N*-formylmethionyl peptides from bacteria. Neutrophils contain lysozyme and alkaline phosphatase within their granules. They die soon after phagocytosing bacteria and are added to the pus, which consists of dead neutrophils, serum, and tissue fluids.

Eosinophils have less phagocytic ability than neutrophils and may kill parasites by either phagocytosis or exocytotic release of granules. Eosinophils contain major basic protein, histaminase, acid phosphatase, and other lysosomal enzymes. They are essential for the destruction of parasites such as trichinae and schistosomes. Mast cells and basophils appear to be involved in the very early stages of parasitic infection, and they are involved in the attraction of eosinophils to the site of infection. This also occurs in nonparasitic infections and involves at least two chemoattractants: histamine and eosinophil-chemoattractant factor of anaphylaxis (ECF-A). Eosinophils inhibit the activity of slow-reacting substance of anaphylaxis (SRS-A), also called *leukotriene 3,* a vasoconstrictive substance produced by basophils and mast cells that also increases vascular permeability.

Basophils are also phagocytic granulocytes but are involved in inflammation through the release of histamine and heparin. Immunoglobulin E produced by plasma cells becomes bound to the cell surface of mast cells and basophils on first exposure. At the time of secondary exposure, the antigen binds to the IgE and stimulates the degranulation of mast cell and basophil granules that contain histamine and heparin. Basophils and mast cells are involved in anaphylactic and immediate hypersensitivity reactions. Delayed-type hypersensitivity is modulated by basophils.

Monocytes are the precursors of macrophages. Microglia, Kupffer cells, and Langerhans cells are macrophages of the brain, liver, and skin, respectively. Unlike neutrophils, eosinophils, and basophils, monocytes are agranular.

Plasma cells differentiate from B cells. B cells are lymphocytes that undergo mitosis and form a plasma cell and a memory cell after exposure to appropriate antigen. An antigen-presenting cell and a specific subtype of T lymphocyte called a *helper T cell* are required for B-cell differentiation into antibody-producing plasma cells.

283–287. The answers are 283-A, 284-E, 285-D, 286-F, 287-B. *(Junqueira, 8/e, pp 222–232, 243, 247. Ross, 3/e, pp 210–213.)* The basophil (B) is about the same size as the neutrophil (A) and contains granules of variable size that may cover the nucleus. The nucleus of the basophil is irregularly lobed with condensed chromatin. The eosinophil (C) is bilobed with more regular granules than the basophil. The monocyte (D) contains an eccentric nucleus, which is often kidney-shaped. The chromatin generally has a ropy appearance and therefore is less condensed than the chromatin of the lymphocyte. The megakaryocyte (E) is a large cell with a multilobular appearance and is the source of platelets. The lymphocyte (F) is considered an agranular cell with an ovoid nucleus. The shape and the arrangement of chromatin vary depending upon the classification of the lymphocyte—small, medium, or large.

Lymphoid System and Cellular Immunology

DIRECTIONS: Each question below contains five suggested responses. Select the **one best** response to each question.

288. Clonal selection functions to

(A) increase the antigen specificity of the immune system
(B) stimulate immunoglobulin class switching
(C) stimulate the production of self-reacting lymphocytes
(D) form specific colony forming units for erythropoiesis and granulopoiesis in the bone marrow
(E) choose the appropriate homing receptors for lymphocytes

289. Antigen-specific cell-mediated immunity as occurs in graft rejection results directly from the activity of

(A) T lymphocytes
(B) plasma cells
(C) monocytes
(D) eosinophils
(E) mast cells

290. Anaphylactic shock is primarily due to the action of

(A) macrophages
(B) mast cells
(C) T lymphocytes
(D) B lymphocytes
(E) eosinophils

291. Mast cells, upon second exposure to a specific antigen, can trigger an anaphylactic reaction by

(A) releasing the content of granules following ingestion of opsonin-coated particles
(B) binding to plasma cells through IgE receptors on the mast cells
(C) causing the histamine-mediated contraction of smooth muscle, as in bronchioles
(D) releasing IgE from its receptors after binding of the antigen to IgE
(E) decreasing the permeability of the capillaries

Questions 292–293

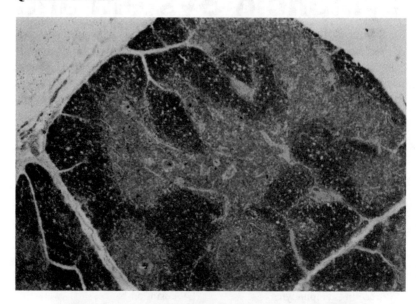

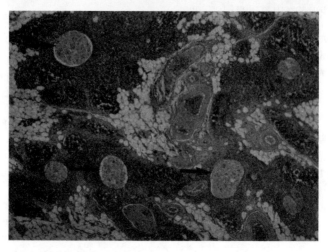

292. This organ, shown at low (*top*) and high (*bottom*) magnification, carries out

(A) generation of self/nonself discrimination
(B) production of antibodies
(C) maturation of natural killer cells
(D) phagocytosis of bacteria
(E) production of memory B cells

293. The structures labeled with the arrows are

(A) trabecular veins
(B) trabecular arteries
(C) tonsillar arteries
(D) Hassall's corpuscles
(E) medullary sinuses

294. Gene rearrangement of cytotoxic T cells occurs primarily in the

(A) bone marrow
(B) spleen
(C) germinal centers
(D) thymus
(E) mesenteric lymph nodes

295. Gene products of class II major histocompatibility complex (MHC) present antigenic peptides to

(A) helper T cells
(B) cytotoxic T cells
(C) antigen-presenting cells
(D) B cells
(E) plasma cells

296. Expression of antigen associated with class I major histocompatibility complex (MHC) molecules is recognized primarily by

(A) B cells
(B) CD4+ cells
(C) CD8+ cells
(D) plasma cells
(E) macrophages

297. Interleukin 2 is made by

(A) plasma cells
(B) natural killer cells
(C) CD4+ T lymphocytes
(D) CD8+ T lymphocytes
(E) macrophages

Questions 298–302

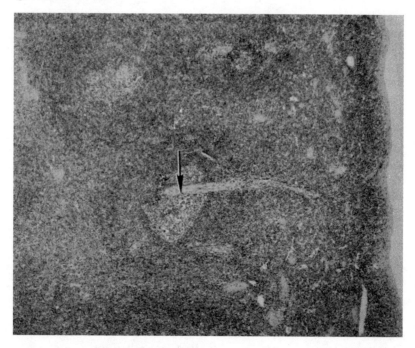

298. The tissue immediately surrounding the structure labeled with the arrow functions primarily as a

(A) T-cell dependent area
(B) B-cell dependent area
(C) region specialized for phagocytosis
(D) region specialized for T-cell education
(E) region specialized for white cell destruction

299. A cell in the structure labeled with the arrow would have already traveled through which of the following?

(A) Capsular arteries of the thymus
(B) Venous sinusoids
(C) Hilar arteries of the lymph node
(D) Trabecular arteries
(E) Trabecular veins

300. This photomicrograph is a medium-magnification view of

(A) lymph node
(B) tonsil
(C) thymus
(D) bone marrow
(E) spleen

301. The structure labeled with the arrow is

(A) Hassall's corpuscle
(B) central vein
(C) central artery
(D) afferent lymphatic vessel
(E) venous sinusoid

302. The structure labeled with the arrow is associated with the

(A) red pulp
(B) white pulp
(C) deep cortex
(D) medullary sinuses
(E) subcapsular sinus

303. Which of the following statements accurately describes the thymus?

(A) It is derived embryologically from the third and fourth branchial arches
(B) It produces CD4+ CD8+ (double positive) cells that emigrate from the thymus and seed to T-dependent areas
(C) It produces single positive (CD4+ or CD8+) cells with appropriate homing receptors and T-cell receptors
(D) It is a secondary lymphoid organ with germinal centers
(E) It develops late in gestation compared with other lymphoid organs

304. Which of the following statements regarding the circulation and recirculation of lymphocytes is true?

(A) Homing receptors on lymphocytes recognize vascular addressins on high endothelial venule cells
(B) Lymphocytes leave the lymph node by diapedesis through the wall of the postcapillary venule to enter the bloodstream
(C) These processes occur only at times of systemic infection
(D) Plasma cells enter the bloodstream from the medulla of the lymph nodes and circulate through the body
(E) High endothelial venules are found only in the lymph nodes

305. The cell shown in the accompanying photomicrograph is treated with fluorescein-labeled antihuman immunoglobulin. The positive cell-membrane fluorescence indicates that the cell is a

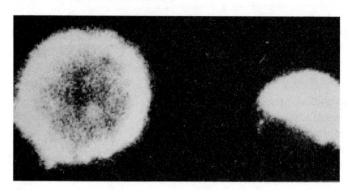

(A) B cell
(B) helper T cell
(C) cytotoxic T cell
(D) transformed T cell
(E) thymocyte

---

306. Gene rearrangement of immunoglobulin occurs primarily in the

(A) bone marrow
(B) spleen
(C) tonsils
(D) thymus
(E) lymph nodes

307. The process known as *immunoglobulin switching* (IgM to IgG) occurs primarily in the

(A) bone marrow
(B) peripheral blood
(C) germinal centers
(D) thymus
(E) splenic red pulp

308. Which of the following allows B cells to recognize antigen?

(A) Gene rearrangement of immunoglobulin
(B) Interaction with antigen in the fetal bone marrow
(C) Expression of IgM molecules on the B cell membrane
(D) Production of IL-4 by helper T cells
(E) Expression of gene products of major histocompatibility complex (MHC)

DIRECTIONS: Each numbered question or incomplete statement below is NEGATIVELY phrased. Select the **one best** lettered response.

309. Education of T cells in the thymus includes all the following EXCEPT

(A) "exit visas" in the form of specific cell surface markers
(B) development of T-cell receptors for antigen
(C) selection of an appropriate subset marker (i.e., CD4+ or CD8+)
(D) expression of appropriate lymphocyte homing receptors
(E) expression of elevated levels of specific integrins (i.e., lymphocyte functional antigen)

310. In DiGeorge syndrome, the third and fourth branchial pouches fail to form. The result is the absence of the parathyroid glands and the thymus. This syndrome can be mimicked by experimental thymectomy. After early thymectomy, one would expect to observe all the following EXCEPT

(A) reduced numbers of plasma cells in the follicular areas of the spleen
(B) reduced numbers of cells in the paracortex (deep cortex) of the lymph nodes
(C) reduced numbers of cells in the periarteriolar lymphoid sheath of the spleen
(D) reduced antibody production in the medullae of the lymph nodes
(E) absence of B cells in the tonsils

311. In a positive tuberculin skin test, helper T cells help in all the following ways EXCEPT

(A) autocrine-mediated increase in proliferation of helper T cells by IL-2
(B) the up-regulation of IL-2 receptors on helper T cells
(C) secretion of IL-4, IL-5, and IL-6, which promote B cell proliferation
(D) secretion of IL-1
(E) attraction and activation of macrophages by release of gamma-interferon

312. All the following will occur during viral infection EXCEPT

(A) macrophages may phagocytose virus
(B) CD4+ T cells respond to viral antigens and MHC class I molecules to directly attack virus-infected cells
(C) large lymphocytes divide to form plasma cells and memory B cells or cytotoxic T cells and memory T cells
(D) T- and B-cell areas of the spleen and lymph nodes are involved in the filtration of the blood and lymph, respectively
(E) B cells in the presence of helper T cells and antigen-presenting macrophages differentiate into plasma cells

313. The photomicrograph below is a histologic section of a lymph node. All the following would be found in the region marked with the asterisk EXCEPT

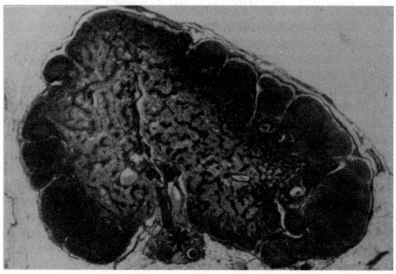

Courtesy of Dr. Vincent H. Guttone, II

(A) efferent lymphatic
(B) afferent lymphatics
(C) vein
(D) artery
(E) convergence of the medullary sinuses

---

314. Macrophages are directly involved in immune responses in all the following ways EXCEPT

(A) production of interleukin 1 (IL-1)
(B) presentation of antigens
(C) killing of tumor cells
(D) production of antibodies
(E) phagocytosis

315. In comparison to the primary immune response, all the following are true of the secondary response EXCEPT

(A) it has a shorter lag period
(B) it has a longer duration
(C) it is of greater intensity
(D) it lacks specificity
(E) it is due to the presence of memory B and T cells

DIRECTIONS: The group of questions below consists of lettered headings followed by a set of numbered items. For each numbered item select the **one** lettered heading with which it is **most** closely associated. Each lettered heading may be used **once, more than once, or not at all.**

Questions 316–319

Match each description with the correct immunoglobulin.

(A) IgG
(B) IgA
(C) IgM
(D) IgE
(E) IgD

316. Is responsible for secretory immunity

317. Is the only immunoglobulin that crosses the placenta

318. Has great affinity for receptors on basophils and mast cells

319. Is found with IgM on the surface of B lymphocytes

Lymphoid System and Cellular Immunology
Answers

288. The answer is A. *(Alberts, 3/e, pp 1199–1200. Roitt, 3/e, pp 1.8–1.9, 2.10.)* Clonal selection is the method by which B and T cells develop specificity for antigens they have not yet seen. B cells function through humoral immunity (i.e., secretion of antibodies specific for the antigen that generated them); T cells are involved in cell-mediated immunity (e.g., the cytotoxic action of CD8+ T cells). In clonal selection a clone of lymphocytes is committed to respond to a particular antigen. The antigenic determinants, which consist of specific amino acids or monosaccharides, actually induce many clones and a wide variety of humoral and cell-mediated responses. This occurs during the development and maturation of the immune system and is responsible for the specificity of lymphocyte cell surface receptors for antigens.

Immunoglobulin class switching occurs during the education (maturation) of B cells. Homing receptor differentiation is an important part of T- and B-cell education in the thymus and bone marrow, respectively.

289. The answer is A. *(Male, 2/e, pp 12, 94. Roitt, 3/e, pp 23.4–23.6.)* Transplant (graft) rejection is mediated by the action of T cells. Helper T (T_H) cells recognize peptides associated with MHC antigens from the donor tissue and become activated. Activated T_H cells release interleukin 2 (IL-2) and interferon-gamma (IFN-γ), which activate cytotoxic T (T_C) cells, B cells, and macrophages. While all of these cells are involved in the rejection process, T cells are the primary agents of transplant rejection. This has been confirmed by animal studies: animals deficient in T cells are incapable of rejecting grafts.

290. The answer is B. *(Isselbacher, 13/e, pp 1630–1632. Junqueira, 8/e, p 107. Male, 2/e, pp 12, 28, 94. Roitt, 3/e, pp 2.18–2.19, 19.2–19.3. Ross, 3/e, p 110.)* Anaphylactic shock is a type of hypersensitivity reaction. In this allergic response, a person who has been sensitized to a particular antigen on first exposure responds with release of secretions from both mast cells and basophils, resulting in smooth muscle contraction (e.g., constriction of bronchioles), increased vascular permeability, and a reduction in blood pressure. In severe cases, circulatory or respiratory failure may occur. The basis of the response is the release of mast cell and basophil secretions on second exposure to an antigen.

291. The answer is C. *(Alberts, 3/e, pp 1208–1211. Junqueira, 8/e, pp 106–107, 249–250.)* Mast cells as well as basophils contain heparin and histamine, which are released by exocytosis. On an initial exposure to an antigen, IgE binds to receptors on the mast cell and basophil surfaces. On second exposure, this bound IgE functions as an antigen receptor. The presence of antigen-antibody complexes on the cell surface induces release of secretion, including release of eosinophil-chemoattractant factor of anaphylaxis (ECF-A), histamine, heparin, and slow-reacting substance of anaphylaxis (SRS-A). In the presence of allergens, many of the symptoms of the allergy are induced by histamine and heparin, which increase vascular permeability and dilate the blood vessels.

292–293. The answers are 292-A, 293-D. *(Male, 2/e, p 20. Roitt, 3/e, pp 2.1, 2.5–2.6, 2.8–2.10, 2.17–2.18, 3.1–3.2.)* The organ in the photomicrograph can be identified at low magnification *(top)* from the lobulation with cortex and medulla in each lobule and the absence of germinal centers. At high magnification *(bottom)* the presence of Hassall's (thymic) corpuscles is an identifying characteristic. Hassall's corpuscles contain degenerating epithelial cells in concentric arrays and increase with age. Their precise function is unknown. The organ in the photomicrograph is the thymus. The thymus functions in the generation of self/nonself discrimination since self-reactive T cells are deleted and self-MHC-restricted cells are expanded during education/maturation within the thymus. Natural killer cells appear to arise in both the bone marrow and thymus, have the morphology of large granular lymphocytes, and mature in the secondary lymphoid organs. Phagocytosis of bacteria is primarily a function of neutrophils (also known as *polymorphonuclear leukocytes*), which arise in the bone marrow. Production of memory B cells as well as effector and memory T cells occurs in the secondary lymphoid organs. Production of antibodies is the responsibility of plasma cells, which arise from B lymphocytes.

294. The answer is D. *(Alberts, 3/e, pp 1228–1229. Roitt, 3/e, pp 2.5–2.7.)* T cells are educated within the thymus during fetal and early neonatal development. There are three types of T cells: cytotoxic T cells, suppressor T cells, and helper T cells. Helper T cells possess a specific cell membrane marker and are known as CD4+ cells, while CD8+ cells have a different cell surface marker and include the suppressor and cytotoxic subgroups. T cells require close contact with other cells in order to carry out their cell-mediated function. This is quite different from B cells, where antibodies are secreted into the bloodstream. The T-cell receptor is composed of α- and β-chains. Each chain contains a variable amino-terminal portion and a constant carboxyl-terminal portion. These chains are encoded for by V, D, J, and C gene segments, which undergo rearrangement during development in the thymus.

295. The answer is A. *(Alberts, 3/e, pp 1238–1240. Junqueira, 8/e, pp 229–230, 250–251. Stevens, p 85.)* Activation of helper T cells is required as an early step in the immune response. Fragments of antigen associated with class II MHC glycoproteins are recognized by helper T cells. The primary cell type for expression of class II MHC is the macrophage, which serves as an antigen-presenting cell, but thymic epithelial cells and B cells can also present antigen under appropriate conditions. In order for B cells to respond to most antigens, helper T cells are an absolute requirement. However, in the case of some bacterial polysaccharides, B cells respond to the antigens in the absence of helper T cells.

296. The answer is C. *(Alberts, 3/e, pp 1235–1238. Stevens, p 85.)* Cytotoxic T cells possess the CD8 cell surface marker. They recognize foreign antigen in association with class I MHC molecules as opposed to helper T cells (CD4+), which recognize antigen in association with class II MHC molecules. Cytotoxic T cells are effective in killing virus-infected cells since these cells express fragments of virus combined with MHC I molecules on their surfaces.

297. The answer is C. *(Junqueira, 8/e, pp 229–230. Stevens, p 83.)* Lymphokines are a subtype of cytokines, secretory products that affect the growth and differentiation of other cells. Interleukin 2 (IL-2) is produced by helper T cells (with the cell surface marker CD4) and stimulates proliferation of T cells.

298–302. The answers are 298-A, 299-D, 300-E, 301-C, 302-B. *(Burkitt, 3/e, pp 214–219. Junqueira, 8/e, pp 260–267. Ross, 3/e, pp 349–353, 365. Stevens, pp 95–98.)* The photomicrograph is a medium-magnification view of the spleen. The histologic structure of the spleen includes the presence of a connective tissue capsule with extensions of the capsule into the parenchyma of the spleen, forming trabeculae. The parenchyma consists of red pulp, which represents areas of red blood cells, many of which are undergoing degradation, and white pulp, which represents lymphocytes involved in the filtration of the blood. The central artery of the white pulp is labeled with the arrow in the photomicrograph. Lymph nodes, in contrast to the spleen, exhibit a distinct cortex and medulla. The tonsils demonstrate an epithelium rather than a connective tissue capsule in their histologic organization. The thymus shows a distinct lobulation and the presence of Hassall's corpuscles as characteristic histólogic features.

The spleen is responsible for the filtration of the blood just as the lymph node serves to filter the lymph. The spleen serves both immunologic and red blood cell functions. The two parts of the spleen—white pulp and red pulp—are dedicated to the two separate functions of the spleen. The spleen plays a minor role in blood storage since the human spleen, unlike that of the dog, is not extensively muscular. The destruction of red blood cells occurs in the red pulp of the spleen. The white pulp of the spleen is involved in the formation of lymphocytes, which occurs during the monitoring of antigen in the bloodstream. The education of B cells occurs in the bone marrow in a similar fashion to the education of T cells, which occurs in the thymus.

The blood supply to the spleen arrives from the aorta, which carries oxygenated blood rich in nutrients. The blood subsequently enters the trabecular arteries, which are surrounded by extensive connective tissue in a spicular form called *trabeculae*. From these arteries blood passes into the central arteries (also known as *white pulp arteries*). These arteries divide into straight (penicillar) arteries and become sheathed by macrophages and lymphoid cells (sometimes known as *sheathed arteries*) before continuing as arterial capillaries that carry blood to the venous sinusoids of the red pulp. From the sinusoids, blood passes into trabecular veins, then the splenic vein, and finally the inferior vena cava.

There has been debate for many years over the nature of the passage of blood from the arterioles into the venous sinuses. Both open (blood entering the tissue spaces first and then filtering into the sinuses) and closed (passage of blood directly from the arterioles into the sinuses of the red pulp) circulations have been proposed. Most experimental, morphological, and clinical data support the concept of at least a partially open circulation with exposure of the blood to macrophages located in the connective tissue surrounding the sinusoids of the red pulp. A blood cell traveling in the central artery would therefore be capable of passing through the capillaries, tissue spaces, venous sinusoids, and trabecular veins, but it would have already traveled through the trabecular arteries on its way to the white pulp.

303. The answer is C. *(Junqueira, 8/e, pp 251–255. Moore, Developing Human, 5/e, pp 195, 199. Paul, 3/e, pp 156–158. Roitt, 3/e, pp 3.7, 18.5. Stevens, pp 86–88.)* The thymus forms from the third and fourth branchial pouches early in development. In a congenital malformation known as DiGeorge syndrome, the thymus and parathyroids fail to develop. The thymus is the first lymphoid organ formed during development, and it functions as a primary lymphoid organ where education of T cells occurs. Cells that emigrate from the thymus are single positive cells (CD4+ or CD8+) with high expression of homing receptors and mature T-cell receptors. Double positive cells as well as cells that recognize self are targeted for destruction by thymic macrophages. The destruction of T cells that recognize self maintains immunologic tolerance. In some diseases—e.g., insulin-dependent diabetes, chronic thyroiditis, and rheumatoid arthritis—T cells recognize self as foreign and destroy β-cells of the pancreas, thyroid parenchyma, and synovial membranes, respectively. The T cells that are educated in the thymus are targeted to secondary lymphoid organs. The maturation of homing receptors on the surface of T cells allows the homing from primary to secondary lymphoid organ.

304. The answer is A. *(Junqueira, 8/e, pp 257–260. Paul, 3/e, pp 179–184. Ross, 3/e, pp 349–353, 364–367. Stevens, pp 91, 100.)* The circulation and recirculation of lymphocytes is a constant process that allows lymphocytes to continuously monitor the presence of antigen. The circulation process also allows augmentation of the immune response to infection. Lymphocytes have specific homing receptors on their cell surfaces that provide entry for mucosal (versus lymph node) seeding. High endothelial venules (HEVs) provide a mechanism for lymphocytes to leave the bloodstream and enter specific areas of the lymph nodes. HEVs are also found in Peyer's patches and during inflammation of tissues (e.g., the synovium in rheumatoid arthritis). T cells home to T-dependent areas of the lymph nodes, spleen, and Peyer's patches. Homing receptors on the lymphocytes are complementary to addressins on the cells of the HEVs. The presence of specific homing receptors on lymphocytes and addressins on the HEVs explains a possible mechanism for the targeting of Peyer's patch–derived lymphocytes to the lamina propria of the gut. The cells that line the HEVs permit the selective passage of lymphocytes by diapedesis through the intercellular junctions. Under normal conditions, HEVs are found in the T-dependent areas of the lymph node, i.e., the deep cortex of the lymph nodes and the interfollicular regions of the Peyer's patches. Plasma cells never enter the bloodstream under normal conditions, but secrete antibodies into the circulation from the medulla of the lymph nodes or the marginal zone of the spleen.

305. The answer is A. *(Roitt, 3/e, p 2.8. Stevens, p 82.)* Lymphocytes that are the equivalent of the bursa-dependent B lymphocytes of the chicken can be identified by the presence of immunoglobulin on their surface membranes. These are the cells that ultimately differentiate into antibody-secreting plasma cells under the appropriate conditions. T lymphocytes, on the other hand, do not have readily detectable cell membrane immunoglobulin.

306. The answer is A. *(Alberts, 3/e, pp 1221–1227. Paul, 3/e, pp 53–54. Stevens, p 82.)* Gene rearrangement of immunoglobulin is the method for generation of antibody diversity and therefore the diversity of the humoral response. This process occurs developmentally during the education of B cells and commits a B cell and its progeny to a genotype. It occurs in the bone marrow of man and other mammals, but in the bursa of Fabricius of birds.

307. The answer is C. *(Alberts, 3/e, pp 1226–1227. Roitt, 3/e, pp 7.14–7.15, 11.13–11.15.)* Immunoglobulin switching normally occurs during the development and maturation of B cells. Synthesis of B-cell antibody begins as IgM inserted into the cell membrane, then switches to membrane-bound IgM and IgD. After antigen stimulation, a switch to surface IgM, IgA, IgG, or IgE occurs, and these antibodies are secreted. Most antibody production occurs in the germinal centers of the lymph nodes, tonsils, and spleen. It occurs in the bone marrow, but the bone marrow also functions in the education of B cells as well as represents the major site of hematopoiesis in the adult. The thymus is responsible for the education of T cells. The splenic red pulp is the site of red blood cell breakdown.

308. The answer is C. *(Alberts, 3/e, pp 1208–1212.)* Recognition of antigen by B cells is accomplished by the expression of IgM molecules on the cell surface. Some investigators use the term *pre-B cell,* or *virgin B cell,* to distinguish those B cells that have not yet synthesized IgM from those that have synthesized and inserted IgM into their cell membranes. IgD, which is produced later by virgin B cells, also serves as an antigen receptor. Through clonal selection, B cells develop specifically for antigens to which they have not been exposed. IL-4 promotes B-cell proliferation, not recognition of antigen. Expression of MHC gene products is essential for T-cell function. Cytotoxic T cells recognize viral protein fragments that have been degraded by a virus-infected cell. Helper T cells recognize class II MHC on the surface of antigen-presenting cells, such as Langerhans cells in the skin and other macrophages in the lymph nodes and spleen.

309. The answer is E. *(Male, 2/e, p 20. Paul, 3/e, pp 156–158. Roitt, 3/e, pp 3.1–3.2, 9.9–9.11, 11.5 Ross, 3/e, pp 346, 368–369.)* As a primary lymphoid organ, the function of the thymus is the education of T cells. This process includes the development of appropriate cell surface markers (e.g., changes in glycoconjugates on the plasma membrane), maturation of the T-cell receptor, and maturation of lymphocyte homing receptors to direct T cells to T-dependent areas of the spleen, lymph nodes, and Peyer's patches. In addition, T cells that emigrate from the thymus express either the CD4+ or the CD8+ subset cell surface marker. Expression of lymphocyte functional antigens (LFA-1 or LFA-2) occurs at the time of lymphocyte activation and enhances binding to other cells and extracellular matrix. Enhancement of integrin expression would inhibit migration, which is required for the emigration and eventual seeding of T cells from the thymus to T-dependent areas.

310. The answer is E. *(Isselbacher, 13/e, p 1163. Junqueira, 8/e, pp 255, 259, 264. Stevens, pp 94, 97, 100.)* The thymus and the bone marrow are the two primary lymphoid organs. While the bone marrow is the site of education of B cells, the thymus is the site of education of T cells. Pre-T cells enter the thymus and differentiate into single positive cells with mature homing and T-cell receptors. DiGeorge syndrome results in the absence of the thymus. In this syndrome there is a failure in the education of T cells and the eventual seeding of T cells to the secondary lymphoid organs. T-dependent areas are found in the lymph nodes (paracortex, or deep cortex), spleen (periarteriolar lymphoid sheath), and the interfollicular regions of the Peyer's patches. Germinal centers will still occur after early thymectomy; however, the absence of helper T cells from the germinal centers will decrease antibody production.

311. The answer is D. *(Alberts, 3/e, pp 1242–1245. Cotran, 5/e, pp 187–189. Isselbacher, 13/e, pp 714–715, 1557–1558. Junqueira, 8/e, pp 229–230, 250–251. Stevens, pp 83–85.)* In a tuberculin skin test, an extract of tuberculin (an antigen of lipoprotein composition obtained from the tubercle bacillus) is injected into the skin of a person who has had tuberculosis or has been immunized against tuberculosis. Memory helper T cells react to the tuberculin and secrete IL-2, which up-regulates IL-2 receptors. IL-2 binding to IL-2 receptors on the same cell is an example of autocrine regulation in which a cell secretes a ligand for a receptor on its own surface. The result of this up-regulation and ligand-receptor binding is an increase in T-cell proliferation. Helper T cells also secrete IL-4, IL-5, and IL-6, which stimulate B-cell proliferation. The production of gamma-interferon by helper T cells results in the attraction and activation of macrophages to the site. Gamma-inteferon also converts other cells (such as endothelial cells) to antigen-presenting cells by induction of class II MHC expression, which further augments the response. The result of the activity of helper T cells is a dramatic increase in the number of lymphocytes and macrophages at the test site, which produces swelling. IL-1 is only synthesized by antigen-presenting cells and helper T cells are the targets.

312. The answer is B. *(Alberts, 3/e, pp 1197, 1202–1204, 1238–1240. Ross, 3/e, p 334.)* During a viral infection, both cell-mediated and humoral responses are stimulated. In these responses, macrophages phagocytose virus. Cells that become infected with virus can be killed by CD8+ cytotoxic T cells, which can react to the antigen in the presence of MHC class I molecules. T- and B-cell areas of the spleen and lymph nodes will be involved in the filtration of the blood and lymph, respectively. B-cell differentiation requires the presence of CD4+ helper T cells and an antigen-presenting cell. The antigen-presenting cell will phagocytose the virus and present it to helper T cells in the presence of MHC class II molecules. The B cell also presents antigen in this arrangement. It will form a plasma cell and a memory B cell. Activated T cells also enlarge to form large lymphocytes and subsequently undergo cell proliferation to form T cells and memory T cells.

313. The answer is B. *(Burkitt, 3/e, pp 203–210. Ross, 3/e, pp 356–359.)* In the histologic section of a lymph node, there is a distinctive cortex and medulla with a connective tissue capsule. The organ possesses the classical bean-shape with a hilum (marked by an asterisk in the figure). Afferent lymphatics enter the lymph node on the convex side and lymph percolates through the subcapsular, cortical, and medullary sinuses. The medullary sinuses converge on the hilum, where the efferent lymphatic vessel drains the node. The hilum also contains an artery and a vein.

314. The answer is D. *(Alberts, 3/e, pp 1238–1240. Junqueira, 8/e, pp 250–251. Stevens, p 85.)* Macrophages are a group of monocyte-derived phagocytic cells that arise from bone marrow and include the Kupffer cells of the liver, Langerhans cells of the skin, and microglia of the central nervous system. They are antigen-presenting cells that produce IL-1. Antigen presentation is the process by which macrophages phagocytose antigen and partially degrade the antigen in the endosomal system. Certain portions of the antigen are returned to the cell surface. IL-1 is a lymphokine (i.e., cytokine) that activates the helper T cell. While macrophages are required for the differentiation of plasma cells from B cells, they are not directly involved in antibody production.

315. The answer is D. *(Alberts, 3/e, pp 1202–1204. Junqueira, 8/e, pp 229–230, 250–251. Stevens, pp 83–85.)* Humoral immunity and cell-mediated immunity involve retention of immunologic memory through memory B and T cells, respectively. A secondary immune response may involve memory T cells, helper cells, macrophages, and memory B cells. The proliferation of either T or B cells during the first exposure to antigen results in the production of memory cells. The response of the memory cells, which remember the antigen, is more rapid, of longer duration, and more intense than the primary immune response. Specificity is retained. For example, the introduction of a different (new) antigen induces a primary rather than a secondary response.

316–319. The answers are 316-B, 317-A, 318-D, 319-E. *(Alberts, 3/e, pp 1206–1212. Junqueira, 8/e, pp 247–250. Ross, 3/e, p 336. Stevens, p 82.)* Immunoglobulins, or antibodies, are glycoproteins synthesized by plasma cells and secreted into the blood plasma. They are specific for the one or more antigenic determinants responsible for their generation. Immunoglobulins may be separated into five subtypes, or classes.

IgG is the most common immunoglobulin and is formed from two identical heavy and light chains that interact through disulfide and noncovalent chemical bonds. Fab fragments are the amino-terminal portions of the light and heavy chains that represent the antigen binding sites. Variation in amino acid sequences in this portion of the molecule is responsible for the diversity of antibody structure. The Fc fragments are the carboxyl-terminal portions of the two heavy chains and represent the receptor binding sites.

IgM, a pentamer, is the first antibody produced by a B cell and is the primary immunoglobulin involved in early immune responses. It is usually found with IgD. Together they are found on the cell surface of B lymphocytes, where they function as antigen receptors and are involved in the differentiation of B cells. IgM and IgD are also found in the circulation.

IgA is found primarily in the bodily secretions, such as the milk and saliva, in the form of a dimer. It consists of two IgA molecules linked by a protein called the *J chain*. Secretory IgA is combined with secretory component, which protects IgA from proteolytic enzymes in the body secretions and facilitates its transport.

IgE is a monomer with strong affinity for IgE receptors on the surface of basophils and mast cells. On first exposure to IgE, there is binding to mast cell and basophil receptors. On second exposure to the same antigen, the antigen-antibody complex on the surface of these cells stimulates the release of histamine and heparin. In addition, the release of eosinophil-chemoattractant factor brings eosinophils into the area to counteract the histamine secretion by mast cells and basophils.

Respiratory System

DIRECTIONS: Each question below contains five suggested responses. Select the **one best** response to each question.

320. The bronchioles do not collapse during exhalation because they are

(A) supported by hyaline cartilage
(B) supported by elastic cartilage
(C) supported by the esophagus
(D) tethered to the lung parenchyma by elastin
(E) supported by multiple layers of smooth muscle

321. The smallest active functional unit of the lung is

(A) an alveolus
(B) a respiratory bronchiolar unit
(C) a bronchopulmonary segment
(D) segmental bronchi
(E) an intrapulmonary bronchus

322. The lung cells known as "congestive heart failure cells" are

(A) type I pneumocytes
(B) type II pneumocytes
(C) macrophages
(D) erythrocytes
(E) fibroblasts

323. Cystic fibrosis (CF) is an important genetic, pediatric disorder in which there is a defect in the cystic fibrosis transmembrane conductance regulator (CFTR), a protein that functions as a chloride channel. Abnormalities of CF include

(A) a decreased concentration of chloride in the sweat
(B) increased chloride secretion into the airways
(C) decreased water reabsorption from the lumen of the airways
(D) decreased active sodium absorption
(E) accumulation of mucus in the pancreatic and salivary ducts

324. In the respiratory system, the ciliated epithelial cells

(A) are mainly located in the respiratory portions of the airways
(B) serve to sweep mucus toward the peripheral lung, where it is phagocytosed by alveolar macrophages
(C) are the primary source of mucus
(D) are ineffective in Kartagener's syndrome
(E) are also known as *brush cells*

325. In the olfactory epithelium, the olfactory receptor cells

(A) are unipolar neurons
(B) send dendritic processes to the olfactory bulb
(C) send axonal processes to the surface epithelium in close proximity to the cilia
(D) are associated with odor-specific transmembrane receptors, which reside in the olfactory cilia
(E) have axonal processes that are confined to the epithelial layer

326. Which of the following statements is true of Clara cells?

(A) They are found in the alveoli
(B) They are the ciliated cells of the bronchiole
(C) They are the endocrine cells that form neuroepithelial bodies
(D) They produce a surfactant-associated protein and contain extensive smooth endoplasmic reticulum (SER)
(E) They are involved in the production of mucus

327. Premature infants may develop respiratory distress syndrome mainly because of

(A) an incomplete endothelial lining
(B) insufficient development and maturation of type II pneumocytes
(C) lack of maturation of type I pneumocytes
(D) disrupted alveolar walls
(E) an excess of alveolar macrophages

328. Which of the following statements describes type I pneumocytes?

(A) They are the precursors of the type II pneumocytes
(B) They are the predominant cell in the interalveolar septum (wall)
(C) They have loose intercellular junctions
(D) They are the source of carbonic anhydrase, an enzyme involved in CO_2 exchange in the capillaries
(E) They provide an extremely thin, gas-permeable barrier

329. Surfactant is accurately described by which of the following statements?

(A) It is recycled through capillary endothelial cells
(B) It is associated with specific proteins that are involved in anchorage, endocytosis of surfactant, and activation of macrophages
(C) Production is inhibited by thyroid hormones that prevent the maturation of type II pneumocytes
(D) Production is inhibited by estradiol that prevents the maturation of type II pneumocytes
(E) Production is inhibited by tubulomyelin figures released from lamellar bodies

330. The cell labeled with the arrows in the electron micrograph

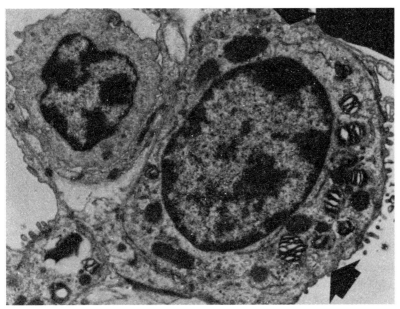

Courtesy of Dr. Kuen-Shan Hung.

(A) is a type I pneumocyte
(B) secretes a substance that increases the surface tension of the pulmonary alveoli
(C) secretes a substance that may be detected by measuring the lecithin/sphingomyelin ratio in the amniotic fluid
(D) is a dust cell, or pulmonary macrophage
(E) is responsible for bone resorption

331. Which of the following procedures is most effective for prevention and treatment of respiratory distress syndrome?

(A) Glucocorticoids given to the premature infant
(B) Isolated surfactant given to the premature infant
(C) Anti-fibroblast-pneumocyte factor given to the premature infant
(D) Testosterone given to the premature infant
(E) Insulin given to the mother during gestation

332. The alveolar pores of Kohn

(A) serve to drain surfactant
(B) serve to equalize the interalveolar pressure
(C) are openings in the pulmonary capillary endothelium
(D) are openings in the bronchiolar walls
(E) are caused by bacterial infection

333. Asthma is a consequence of the release of histamine and heparin from

(A) mast cells, which induces edema and bronchoconstriction
(B) plasma cells, which induces antibody release
(C) eosinophils, which induces proliferation of basophils
(D) epithelial cells, which induces hyposecretion
(E) goblet cells, which induces hypersecretion

DIRECTIONS: Each numbered question or incomplete statement below is NEGATIVELY phrased. Select the **one best** lettered response.

334. Olfactory signal transduction is similar to signal transduction in rod cells stimulated by light. Olfactory signal transduction involves all the following EXCEPT

(A) cyclic AMP–gated ion channels
(B) activation of a G protein that binds GTP
(C) stimulation of cyclic AMP
(D) depolarization of the olfactory receptor cell in the olfactory epithelium, leading to excitation
(E) cyclic GMP activity dependent on protein kinase activation

335. Defense mechanisms of the respiratory system are based on all the following EXCEPT

(A) phagocytic activity of type II pneumocytes
(B) nasal clearance of material
(C) mucociliary action for tracheo-bronchial clearance
(D) phagocytic action of alveolar macrophages
(E) release of immunoglobulins from resident plasma cells

336. All the following belong to the conducting portion of the respiratory system EXCEPT the

(A) trachea
(B) bronchi
(C) bronchioles
(D) alveoli
(E) larynx

337. Emphysema is a common occurrence in persons deficient in alpha$_1$-antitrypsin and in those who smoke. All the following may occur in emphysema EXCEPT

(A) increased levels of elastase
(B) increased numbers of neu-trophils or macrophages
(C) bronchiolar collapse
(D) decreased antiprotease activity
(E) antielastase activity stimulated by smoking

338. All the following are part of the minimal blood-air barrier in the lungs EXCEPT

(A) basal laminae
(B) alveolar endothelial cells
(C) alveolar macrophages
(D) type I pneumocytes
(E) a space of 0.3 to 0.7 μm in thickness

Respiratory System
Answers

320. The answer is B. *(Stevens, p 135.)* The bronchioles, unlike the trachea and bronchi, do not contain hyaline cartilage. The bronchi are intrapulmonary structures and are not associated with the esophagus. A relatively thick layer of smooth muscle is found in the bronchioles. They are tethered to the lung parenchyma by elastic tissue, which also plays a role in the stretch and recoil of the lungs during inhalation and exhalation.

321. The answer is B. *(Moore, Anatomy, 3/e, pp 72–77. Ross, 3/e, pp 540–541.)* The smallest functional unit of the lung is the respiratory bronchiolar unit, which contains a respiratory bronchiole and the alveoli associated with it. This unit allows for conduction and gas exchange. The alveolus is only associated with gas exchange and the bronchi form part of the conduction system. The bronchopulmonary segment is a functional unit of lung structure, but it is not the smallest unit. Bronchopulmonary segments are particularly important in surgical resections of the lung since they represent functional units with connective tissue boundaries and individualized vasculature, including pulmonary and bronchial arteries, pulmonary lymphatics, and pulmonary nerves, all of which follow the air-conducting system of the bronchial tree and its branches.

322. The answer is C. *(Cotran, 5/e, p 522. Junqueira, 8/e, p 341. Ross, 3/e, p 546.)* During congestive heart failure, edema results in leakage of erythrocytes into the alveoli. Transferrin and hemoglobin are also present in the edematous fluid released from the capillaries. These two products are phagocytosed by alveolar macrophages, which convert these products to hemosiderin. The hemosiderin-containing alveolar macrophage has been called the "congestive heart failure cell." The presence of these cells is an indicator of edematous lung changes.

323. The answer is E. *(Cotran, 5/e, pp 451–454. Isselbacher, 13/e, pp 1149–1150.)* Cystic fibrosis is a frequent occurrence in white children (1 in 200 births). It is a genetic disease in which the defect has been determined to occur in the CFTR protein, which functions as a chloride channel. In the sweat glands a decrease in sodium transport results in increased chloride levels in the sweat (the original detection test for CF). In the airways, decreased chloride secretion occurs in conjunction with active sodium absorption, resulting in loss of water from the lumen as water follows sodium. The result is increased viscosity of mucous secretions and obstruction of the airways and other organs. The pancreas and salivary gland secretions are affected in a similar way, although these abnormalities do not occur in some cases. In the case of the lungs, the loss of the mucociliary escalator action results in susceptibility to opportunistic lung infections.

324. The answer is D. *(Junqueira, 8/e, pp 326–330. Ross, 3/e, pp 537–538. Stevens, p 131.)* The respiratory epithelium is pseudostratified and ciliated and comprises a number of different cell types. Ciliated epithelial cells are the primary cell type of the respiratory epithelium. They are mainly located in the conducting portions of the airways, where their cilia sweep mucus upward from the lung toward the pharynx. Kartagener's (immotile cilia) syndrome results in respiratory infections because of a deficiency of dynein in the cilia. The goblet cells produce mucus composed mostly of glycoproteins, which are anionic and attract particles. Mucus serves as the first line of defense for the respiratory system. The brush cells are a third type of respiratory epithelial cell with apical microvilli that may function as sensory receptors. The basal cells of the pseudostratified epithelium proliferate and are responsible for the replacement of the epithelium. The last cell type is the neuroendocrine cells, which are involved in paracrine (local effects on other cells) and endocrine regulation (through blood-borne secretions) of respiratory function.

325. The answer is D. *(Junqueira, 8/e, pp 448–450. Ross, 3/e, pp 531–533. Stevens, p 126.)* The olfactory receptor epithelial cells are bipolar cells that send a dendritic process to the surface epithelium for the reception of olfactory information. There are odor-specific transmembrane receptors present in the olfactory cilia. Axonal processes leave the epithelium and form the olfactory bulb, which contains the first cranial (olfactory) nerve. There appears to be a stem cell pool for replacement of olfactory receptor cells even in adult mammals, which is an exception to the rule that neurons are not replaced in the adult.

326. The answer is D. *(Junqueira, 8/e, p 333. Ross, 3/e, pp 541–543, 556–557. Stevens, p 132.)* The Clara cells are found in the bronchioles. Other cell types in the bronchiolar epithelium include ciliated, goblet, and endocrine cells. The goblet cells make mucus, but the number of goblet cells in the bronchioles is fewer than in the bronchi. Ciliated cells are involved in the sweeping of mucus and attached particles toward the nasal cavities. Endocrine cells develop earlier than other cells in the lungs and may be involved in the regulation of growth and development. Endocrine cells may be located singly but are frequently found in aggregations called *neuroepithelial bodies (NEBs)* within the bronchiolar epithelium. The endocrine cells of the NEBs synthesize serotonin, enkephalin, calcitonin, bombesin, and other peptides. Clara cells have extensive SER, which is typical of steroid-secreting cells. They are known to synthesize both glycosaminoglycans for protection of the bronchiolar epithelium and surfactant-associated proteins.

327. The answer is B. (*Junqueira, 8/e, pp 338–341. Stevens, p 134.*) Type II pneumocytes synthesize surfactant, but differentiation of type II pneumocytes occurs late in gestation. In premature infants there is a deficiency of surfactant because of the immaturity of the type II pneumocytes. The deficiency of surfactant inhibits normal expansion of the alveoli and results in hyaline membrane disease and respiratory distress syndrome. Glucocorticoids may be used to induce surfactant production.

328. The answer is E. (*Junqueira, 8/e, pp 336–337. Stevens, pp 133–134.*) Type I pneumocytes cover 97 percent of the alveolar surfaces. However, they are not the predominant cell type in the alveolar wall since they account for only about 10 percent of the cells in this location. The alveolar septum, or wall, consists primarily of capillary endothelial cells and fibroblasts, although type I and type II pneumocytes are present. Type I and type II pneumocytes originate from the same precursor cell. The type I cells have very tight junctions that inhibit leakage of pulmonary tissue fluid into the alveolar air spaces. These cells are very thin and provide a gas-permeable barrier. Red blood cells are the source of the enzyme carbonic anhydrase, which is involved in CO_2 gas exchange.

329. The answer is B. (*Junqueira, 8/e, pp 337–341. Stevens, p 134.*) Surfactant consists of an aqueous layer, or hypophase, that contains proteins and mucopolysaccharides. This layer is covered by a layer of phospholipid that consists predominantly of dipalmitoyl phosphatidylcholine. This second layer is the functional surfactant layer. The release of lamellar bodies by exocytosis is followed by their general unraveling to form tubulomyelin figures. The tubulomyelin consists of a crisscross lipid bilayer, which covers the type II pneumocytes. There are a number of surfactant-associated proteins involved in the stabilization of the surfactant, activation of surfactant recycling, and enhancement of the surfactant-induced reduction of surface tension. Three mechanisms have been described for turnover of surfactant: (1) receptor-mediated endocytosis by type I pneumocytes, (2) phagocytosis by macrophages, and (3) endocytosis by type II pneumocytes, which reutilize surfactant through the lysosomal system and shuttle components to the lamellar bodies.

330. The answer is C. *(Junqueira, 8/e, pp 337–341. Moore, Developing Human, 5/e, pp 105, 232–233. Ross, 3/e, p 544. Stevens, p 134.)* The cell labeled in the electron micrograph is a type II pneumocyte. Surfactant is produced by type II pneumocytes in the lung and is stored in the form of lamellar bodies, which are observed by electron microscopy. The primary component of the surfactant is dipalmitoyl phosphatidylcholine (dipalmitoyl lecithin); other constituents include glycosaminoglycans and other phospholipids. The production of surfactant is stimulated by the administration of glucocorticoids. The lecithin/sphingomyelin ratio is a test that can be performed on a sample of amniotic fluid obtained by amniocentesis. It is used to determine whether the type II pneumocytes are mature and are synthesizing and secreting surfactant.

331. The answer is B. *(Cotran, 5/e, p 445.)* Respiratory distress syndrome is best treated with surfactant or synthetic surfactant. Glucocorticoids increase surfactant production by stimulating mesenchymal cells to release fibroblast-pneumocyte factor, which stimulates type II cell secretion. Insulin and testosterone inhibit surfactant production.

332. The answer is B. *(Junqueira, 8/e, pp 337, 342. Ross, 3/e, p 546. Stevens, p 135.)* The pores of Kohn provide a mechanism for connection of one alveolus to another. Macrophages travel from alveolus to alveolus through these pores. The pores normally serve to equalize air pressure between alveoli and can, in the disease state, provide collateral circulation of air in the event a bronchiole is blocked. However, they can also provide a passageway for the spread of bacteria.

333. The answer is A. *(Cotran, 5/e, pp 690–691. Junqueira, 8/e, pp 334, 344. Stevens, p 136.)* Mast cells in the bronchioles are stimulated to release histamine and heparin, which induce the contraction of smooth bronchiolar muscle and edema in the wall. If the bronchoconstriction is chronic, the result will be thickening of the bronchiolar musculature. These muscle changes are usually accompanied by the synthesis of viscous mucus, which can obstruct the airway.

334. The answer is E. *(Alberts, 3/e, pp 752–753.)* The response of rod cells to light and of olfactory receptor cells to odor are two examples of signal transduction that bypasses a protein kinase system. In the case of the olfactory epithelium, an odorant molecule binds to an odor-specific transmembrane receptor found on the modified cilia at the apical surface. The binding activates an odorant-specific G protein (G_{olf}), which binds GTP. The resulting dissociation of the α-subunit stimulates adenylate cyclase to produce cyclic AMP. Cyclic AMP directly stimulates the opening of the cation channels on the membrane of the bipolar olfactory receptor cells, leading to Na^+ influx. The resulting change in membrane potential is transmitted from the modified cilia to the olfactory vesicle through the neuron to the basal axon. The axonal process traverses the lamina propria as the olfactory nerve and passes through the cribriform plate of the ethmoid to terminate in the olfactory bulb. In the case of the rod, the cyclic nucleotide involved is cyclic GMP not cyclic AMP. Another difference in the rod is that light causes hyperpolarization rather than depolarization.

335.   The answer is A. *(Cotran, 5/e, p 696. Junqueira, 8/e, p 344.)* The respiratory system is exposed to constant assault from the environment. In order to protect the distal portions of the lung, which under normal conditions are considered a sterile environment, extensive defense mechanisms have evolved. Nasal clearance of material occurs through sneezing, while other material located posteriorly may be swept into the nasopharynx. The mucociliary action within the trachea and bronchi is often called the *mucociliary,* or *tracheobronchial, escalator.* At the distal end of the system, the alveolar macrophages phagocytose foreign material and secrete and respond to an array of cytokines. The type II pneumocytes resorb as well as secrete surfactant; however, this function does not play a major role in host defense mechanisms.

In the bronchi, there is extensive lymphoid tissue, which is known as the *bronchus-associated lymphoid tissue (BALT)* and is analogous to the mucosa-associated lymphoid tissue (MALT) of the gut and the skin-associated lymphoid tissue (SALT). There are B- and T-cell areas throughout the BALT. The B cells are precursors of plasma cells and synthesize immunoglobulins (e.g., IgA secretion by plasma cells associated with the bronchial glands). T cells are of the same three categories found elsewhere in the body: helper T cells, cytotoxic T cells, and suppressor T cells. Helper T cells recognize foreign antigen in association with class II major histocompatibility complex (MHC) molecules. Cytotoxic T cells recognize fragments of antigen (specifically viral fragments) on the surface of viral-infected cells in association with class I MHC. Antigen-presenting cells (i.e., alveolar macrophages) also function in a similar fashion to those found elsewhere in the body; they present antigen to helper T cells in conjunction with class II MHC.

336. The answer is D. *(Junqueira, 8/e, pp 325–326. Ross, 3/e, p 530. Stevens, pp 124–125.)* The respiratory system is usually divided into two parts: the conducting portion and the respiratory portion. The conducting portion consists of the nasopharynx, larynx, trachea, bronchi, bronchioles, and terminal bronchioles. This portion is a conduit for air to travel to the alveoli, where exchange of oxygen and carbon dioxide occurs. The conducting portion of the respiratory system also conditions the air, which involves moistening, cleansing, and warming the inspired air. The serous glands of the conducting portion of the respiratory system add secretions to the air, which serve to moisten it. The mucous glands add mucus to the air, which facilitates the entrapment of particulate matter. The nasal hairs and cilia lining the conducting portion serve to cleanse the air, while the extensive vasculature in the wall of the conducting portion serves to warm the air as it passes from the nares to the terminal bronchioles. The nasal cavity and paranasal sinuses are lined by a ciliated, mucus-secreting respiratory epithelium that facilitates the cleansing, warming, and moistening functions of the conducting portion.

The respiratory portion of the system is the location of exchange of oxygen and carbon dioxide. Included in this portion are the respiratory bronchioles, the alveolar ducts, and the alveoli.

337. The answer is E. *(Cotran, 5/e, pp 585–587.)* There are genetic and environmental causes of emphysema. The environmental causes include smoking and air pollution, while deficiency in alpha$_1$-antitrypsin (antiprotease) activity is the genetic cause of the disease. In both cases, the balance between normal elastase-elastin production seems to be involved and forms the basis of the protease-antiprotease hypothesis of emphysema. Persons with a deficiency in alpha$_1$-antitrypsin activity lack sufficient antiprotease activity to counteract neutrophil-derived elastase. When there is an increase in the entry and activation of neutrophils in the alveolar space, more elastase is released, and elastic structures are destroyed. In smoking there appears to be an increase in the number of neutrophils and macrophages in alveoli and increased elastase activity from neutrophils and macrophages, coupled with a decrease in antielastase activity because of oxidants in cigarette smoke and antioxidants released from the increased numbers of neutrophils. The increased protease activity causes breakdown of the alveolar walls and dissolution of elastin in the bronchiolar walls. The loss of tethering of the bronchiole to the lung parenchyma leads to collapse of the bronchioles.

338. The answer is C. *(Junqueira, 8/e, pp 335–339. Ross, 3/e, p 546. Stevens, p 135.)* Pulmonary capillaries are sometimes in direct contact with the alveolar wall and in other locations the alveolar wall and capillaries are separated by cells and extracellular fibers. The areas of direct contact are the location of gas exchange while the other areas represent sites of fluid exchange between the interstitium and air spaces. Oxygen moving from the alveolar air to the capillary blood and carbon dioxide diffusing in the opposite direction pass through a three-component blood-air barrier. This barrier consists of type I pneumocytes, endothelial cells, and their fused basal laminae. Macrophages are present for the phagocytosis of debris and surfactant.

Integumentary System

DIRECTIONS: Each question below contains five suggested responses. Select the **one best** response to each question.

Questions 339–340

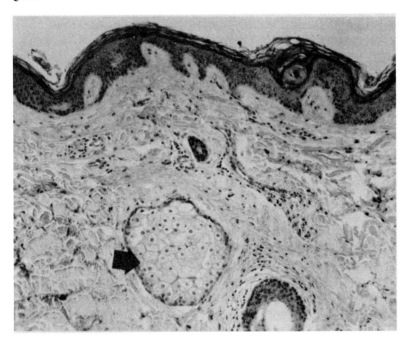

339. The structure indicated by the arrow is a

(A) sweat gland
(B) serous gland
(C) sebaceous gland
(D) hair follicle
(E) nerve

340. The organ in the photomicrograph is

(A) thick skin
(B) thin skin
(C) anus
(D) esophagus
(E) urinary bladder

341. Epidermal Langerhans cells are correctly described by which of the following statements?

(A) They are antigen-presenting cells
(B) They are specialized keratinocytes
(C) They are neuroendocrine cells
(D) They synthesize melanin
(E) They form extensive desmosomes

342. Melanocytes are derived embryologically from

(A) neural crest
(B) neuroectoderm
(C) surface ectoderm
(D) mesoderm
(E) endoderm

343. Merkel cells are modified epidermal cells that function primarily in

(A) cytokine secretion
(B) phagocytosis
(C) antigen presentation
(D) cutaneous sensation
(E) expression of Fc, Ia, and C3 receptors

344. The hair is formed

(A) by keratinization
(B) by phagocytosis of epidermis
(C) by fibroblast collagen synthesis
(D) without the influence of hormones
(E) in a continuous, synchronous fashion

DIRECTIONS: Each numbered question or incomplete statement below is NEGATIVELY phrased. Select the **one best** lettered response.

345. The skin carries out all the following functions EXCEPT

(A) provision of a barrier
(B) endocytosis
(C) excretion
(D) secretion
(E) sensory reception

346. In comparison with thin skin, thick skin has all the following characteristics EXCEPT

(A) a lack of hair follicles
(B) a lack of eccrine glands
(C) a stratum lucidum
(D) a more extensive stratum corneum
(E) a well-developed rete ridge system

347. In melanocytes, all the following events occur EXCEPT

(A) melanin granules are derived from the Golgi
(B) melanin granules are transferred intact to keratinocytes
(C) tyrosinase catalyzes the production of melanin from DOPA
(D) transfer of melanin granules occurs by phagocytosis
(E) melanin granules are degraded by keratinocytes in the skin of whites

348. All the following occur after exposure to ultraviolet (UV) radiation EXCEPT

(A) protection of the nuclei of dividing keratinocytes by melanin
(B) darkening of melanin
(C) increased melanin synthesis
(D) proliferation of melanocytes
(E) increased cytocrine secretion of melanin

349. Epidermal repair following a wound involves all the following EXCEPT

(A) proliferation of basal keratinocytes
(B) proliferation of hair follicles and sweat gland cells
(C) production of extracellular matrix
(D) reepithelialization by differentiation from dermal fibroblasts
(E) proliferation of dermal fibroblasts

DIRECTIONS: Each group of questions below consists of lettered headings followed by a set of numbered items. For each numbered item select the **one** lettered heading with which it is **most** closely associated. Each lettered heading may be used **once, more than once, or not at all.**

Questions 350–353

Match each description with the correct layer of the skin.

(A) Stratum corneum
(B) Stratum basale
(C) Stratum lucidum
(D) Stratum granulosum
(E) Stratum spinosum

350. Presence of hemidesmosomes

351. Site of most mitotic activity in the integument

352. Presence of keratohyalin granules

353. Distinguishing feature of thick skin

Questions 354–357

Match each description with the appropriate type of sensory receptor.

(A) Ruffini endings
(B) Pacinian corpuscle
(C) Meissner corpuscle
(D) Merkel corpuscle
(E) Free nerve endings

354. Perception of deep pressure

355. Perception of pain

356. Perception of fine touch in glabrous skin

357. Attachment to keratinocytes by desmosomes

Integumentary System
Answers

339–340. The answers are 339-C, 340-B. *(Burkitt, 3/e, pp 162–163, 166–168. Erlandsen, pp 105, 107–108. Junqueira, 8/e, pp 72–74, 355–356. Ross, 3/e, pp 383–390. Stevens, pp 356–359, 363.)* The photomicrograph represents a microscopic section obtained from thin skin. The structure marked with the arrow is a sebaceous gland, which is located in the dermis and associated with the hair follicles. The sebaceous glands are holocrine (i.e., they shed the cell along with the secretory product). Sebum is a lipid product released into a duct that terminates in a hair follicle. The presence of sebaceous glands identifies the section as thin skin. Sebaceous glands are not found in thick skin. Another difference between thick and thin skin is the virtual absence of the stratum lucidum in thin skin.

The sweat glands are of two different types: merocrine glands and apocrine glands. The merocrine glands release their secretion through exocytosis with conservation of membrane. In the anal, areolar, and axillary regions the sweat glands are of the apocrine type and empty into the hair follicles. In apocrine glands the apical part of the cell is released with the secretion.

341. The answer is A. *(Junqueira, 8/e, p 351. Ross, 3/e, pp 378–379. Stevens, p 354.)* The Langerhans cells are antigen-presenting cells that phagocytose epidermal antigens and present them in association with class II MHC molecules to a helper T cell. The helper cell assists a B cell in its differentiation into a plasma cell and a memory cell. The specialized neuroendocrine cells of the epidermis are called *Merkel cells* and are believed to be associated with touch reception. Desmosomes are extensive between cells of the stratum basale and stratum spinosum.

342. The answer is A. *(Junqueira, 8/e, p 349. Ross, 3/e, p 375.)* Melanocytes are derived from melanoblasts, which emigrate from the neural crest. These cells migrate into the skin during embryonic development and form a relationship with a specific number of keratinocytes.

343. The answer is D. *(Junqueira, 8/e, p 351. Ross, 3/e, p 379. Stevens, pp 349, 354–355, 361.)* The Merkel cell is a modified keratinocyte found in areas in which fine tactile sensation is critical, such as the fingertips. Merkel cells are associated with an ending of an unmyelinated fiber, forming a

Merkel corpuscle. Langerhans cells function in phagocytosis, antigen presentation, cytokine production, and expression of Fc, Ia, and C3 receptors.

344. The answer is A. *(Junqueira, 8/e, pp 352–355. Ross, 3/e, pp 382–383. Stevens, pp 356–357.)* The hair is formed by keratinization. This process is similar to that which occurs in the epidermis. Hair formation and growth occurs discontinuously and without synchrony. Growth is described as occurring in patches with periods of growth (anagen), followed by a brief hiatus (catagen), and subsequently a lag period in which atrophy of the hair occurs (telogen). The growth process is influenced by a number of hormones: androgens, as well as adrenocortical and thyroid hormones.

345. The answer is B. *(Junqueira, 8/e, p 346. Ross, 3/e, p 370. Stevens, p 348.)* The skin carries out a number of essential functions. It provides a barrier to entry of materials from the environment. This barrier function is related to the ability of the skin to prevent water loss as well as the thermoregulatory function of the skin. The skin also possesses elasticity, which is required when there is swelling during edema or pregnancy. The secretory function is fulfilled by the production of sweat by the eccrine sweat glands and production of sebum by the holocrine sebaceous glands associated with the hair follicles. Sensory function is carried out by free nerve endings, pacinian corpuscles, Meissner corpuscles, and Ruffini endings. The skin has little absorptive capability except for sunlight, important for vitamin D metabolism, and uptake of lipid-soluble substances.

346. The answer is B. *(Ross, 3/e, pp 370–371, 392–395. Stevens, pp 362–363.)* Thick skin is structured to protect against constant trauma. As found on the soles and palms, thick skin lacks hair follicles and contains an additional layer called the *stratum lucidum,* a more extensive stratum corneum, and a well-developed rete ridge system. The stratum lucidum represents a layer of eosinophilic-staining cells that contain keratin. The extensive rete ridges prevent epidermal shearing forces and the extra keratin increases the barrier function of the skin. Eccrine sweat glands and ducts are found in both thin and thick skin, but are very numerous in the thick skin of the palms and soles as well as the thin skin of the axilla and scalp.

347. The answer is C. *(Junqueira, 8/e, pp 348–351. Ross, 3/e, pp 375–378. Stevens, p 353.)* In melanocytes, melanin is synthesized from tyrosine by the action of tyrosinase forming 3,4-dihydroxyphenylalanine (DOPA). The DOPA is subsequently transformed to melanin. This process occurs in melanosomes (immature granules that contain tyrosinase). Mature granules are transferred to keratinocytes by phagocytosis of part of the melanocyte; this process is called *cytocrine secretion* and occurs in the stratum basale and stratum spinosum (malpighian layer). Melanin granules remain relatively intact in persons with black skin; in whites, there is degradation of the granules by lysosomal systems.

348. The answer is D. *(Junqueira, 8/e, pp 346, 348–350. Ross, 3/e, pp 375–378. Stevens, p 353.)* Melanin is synthesized by melanocytes and is transferred to keratinocytes by cytocrine secretion. It is responsible for the pigmentation of the skin and serves to protect the nuclei of dividing keratinocytes in the stratum basale and stratum spinosum. Darkening of the skin in the presence of UV radiation occurs through the darkening of melanin plus an increase in the synthesis and subsequent transfer of melanin (cytocrine secretion). In general, skin color is determined by the number of melanin granules in the skin and not the number of melanocytes per unit area, which is relatively uniform from region to region and between different races.

349. The answer is D. *(Ross, 3/e, p 390.)* Repair of epidermal wounds requires chemoattraction of macrophages to the wound site and removal of damaged tissue by these infiltrating cells. Repair is mediated by the proliferation of endothelial and smooth muscle cells for the repair of blood vessels and angiogenesis. Proliferation of basal keratinocytes and fibroblasts occurs in small wounds for the repair of the epidermis and dermis, respectively. In deeper wounds, new epithelial cells are obtained from the epithelium of the hair follicles and sweat glands located in the dermis. Reepithelialization is inhibited in wounds, which remove all epithelial cells and require skin grafts to enhance the repair process. In these cases there is only minor wound healing by migration from the margins of the wound.

350–353. The answers are 350-B, 351-B, 352-D, 353-C. *(Junqueira, 8/e, pp 346–348. Ross, 3/e, pp 370–375. Stevens, pp 349–352.)* The skin or integument is composed of an epithelial layer (epidermis) and underlying connective tissue (dermis). The epidermis consists of four to five strata (from the basement membrane to the skin surface): stratum basale, stratum spinosum, stratum granulosum, stratum lucidum, and stratum corneum. The basal layer contains most of the mitotic cells and is attached to the basement membrane with hemidesmosomes. The stratum spinosum contains the prickle cells, which have numerous cytoplasmic tonofilaments and intercellular desmosomes. This layer is often classified with the stratum basale as the malpighian layer. The stratum basale and stratum spinosum both contain mitotic keratinocytes, and these are the two layers that show a hyperproliferative state in psoriasis. In this disease, increased cell proliferation leads to a thickening of the epidermis with a shortening of the epidermal turnover period. Under normal conditions, there is a gradual replacement process in the epidermis with the production of new cells in the stratum basale and stratum spinosum and their migration toward the surface as they gradually differentiate. The stratum granulosum may be five cells thick. The cells contain numerous keratohyalin granules. This layer also produces lamellar granules, which form a lipid bilayer barrier to penetration of substances. The stratum lucidum is a translucent layer typical of thick skin. The stratum corneum contains as many as 20 layers of flattened cells and is filled with keratin.

354–357. The answers are 354-B, 355-E, 356-C, 357-D. *(Guyton, 8/e, pp 520–521. Junqueira, 8/e, pp 447–448. Ross, 3/e, pp 379–382. Stevens, pp 354–355, 361.)* The listed sensory receptors are all found in the skin and subserve a variety of functions. Free nerve endings are unencapsulated receptors that function in the reception of many different modalities. Pain receptors in the skin are all free nerve endings. The Ruffini endings are the simplest encapsulated receptor and are associated with collagen fibers. Mechanical stress results in displacement of the collagen fibers and stimulation of the receptor. The pacinian corpuscle is specialized for deep pressure in areas such as the dermis and internal organs (e.g., the pancreas). Its structure resembles an onion with concentric fluid-filled layers surrounding a centrally placed unmyelinated nerve fiber. Displacement of the layers results in the depolarization of the axon. The Meissner corpuscle is found in glabrous (hairless) skin in areas such as the lips and palms and responds to low-frequency stimuli. The Merkel corpuscle consists of a Merkel cell, a modified keratinocyte specialized for acute sensory perception, and a neuron terminal, which forms a disk apposed to the Merkel cell. Merkel cells are attached to neighboring keratinocytes by desmosomes.

Gastrointestinal Tract and Glands

DIRECTIONS: Each question below contains five suggested responses. Select the **one best** response to each question.

358. The resting parietal cell does not secrete acid because

(A) the Na^+,K^+-ATPase is inserted into the apical membrane
(B) the chloride channel of the apical plasma membrane is closed
(C) the H^+,K^+-ATPase is localized in the tubulovesicular membranes
(D) carbonic anyhdrase is sequestered in tubulovesicles
(E) histamine receptors are uncoupled from their second messengers

359. The primary source of bilirubin is

(A) hepatocyte synthesis
(B) reabsorption from the intestine
(C) breakdown of fatty acids
(D) senescent red blood cells
(E) synthesis by the gallbladder

360. The process of restitution occurs in repair of the stomach following damage. This process occurs through

(A) decreased cell cycle time of crypt cells
(B) increased turnover of mucous neck cells
(C) increased secretion of mucous neck cells in the fundic glands
(D) migration of new cells from the neck region of the fundic glands
(E) movement of uninjured cells to cover damaged areas

361. Enteroendocrine cells secrete their products by

(A) a holocrine mechanism
(B) release into the lumen of the crypt of Lieberkühn in the small intestine
(C) an apocrine mechanism
(D) an exocrine mechanism
(E) a paracrine or endocrine method

362. In regard to the enteroendocrine cells and the cells composing the enteric nervous system of the gut, both types of cells

(A) are derived from neural crest
(B) secrete similar peptides
(C) are essential for the intrinsic rhythmicity of the gut
(D) are turned over rapidly
(E) are found only in the small intestine

363. Which of the following processes is involved in lipid absorption and processing by the enterocyte?

(A) Active transport of glycerol across the apical membrane
(B) Exocytosis of chylomicra at the basolateral membranes
(C) Endocytosis of triglycerides at the apical membrane
(D) Uptake of chylomicra by phagocytosis across the apical membrane
(E) Synthesis of triglycerides in the rough endoplasmic reticulum (RER)

364. Hirschsprung disease and Chagas disease result in disturbance of intestinal motility. The site of this disruption is most likely the

(A) vascular smooth muscle
(B) submucosal (Meissner) plexus
(C) muscularis mucosa
(D) myenteric (Auerbach) plexus
(E) enteroendocrine cells

365. As saliva passes through the duct system, which of the following changes occurs?

(A) Active secretion of Na^+
(B) Secretion of Cl^-
(C) Absorption of HCO_3^-
(D) Secretion of K^+
(E) Absorption of K^+

366. The primary regulator of salivary secretion is

(A) the sympathetic nervous system
(B) the parasympathetic nervous system
(C) aldosterone
(D) cholecystokinin
(E) secretin

367. Which of the following occurs during acid secretion by parietal cells?

(A) The Na^+,K^+-ATPase modulated pump on the lateral membrane accumulates sodium and exports potassium
(B) The proton pump in the basolateral membrane brings potassium into the cell across this membrane
(C) A decrease in blood pH occurs during digestion
(D) HCO_3^- from carbonic acid is released directly into the lumen of the gastric glands by active transport
(E) Carbonic anhydrase generates H^+ and HCO_3^-

368. On the electron micrograph of a hepatocyte below, what are the dark structures indicated by the arrows?

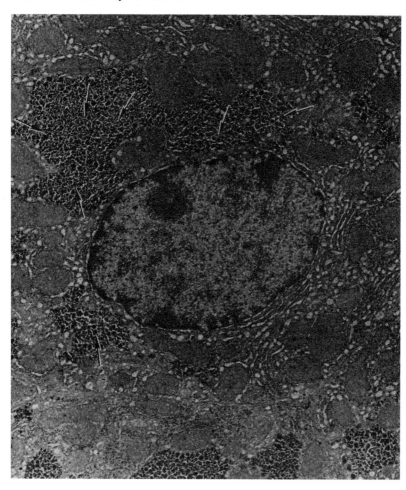

(A) Chylomicra
(B) Glycogen
(C) Mitochondria
(D) Peptide-containing secretory granules
(E) Ribosomes

Questions 369–370

Use the figure below to answer the following two questions. It is a scanning electron micrograph taken from the region between two hepatocytes.

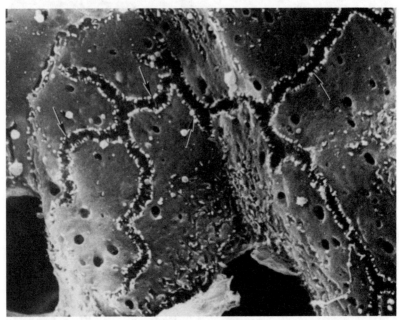

Courtesy of Dr. Kuen-Shan Hung and Karen Grantham, KUMC Electron
Microscopy Center.

369. The branching structures shown in the photomicrograph are

(A) gap junctions
(B) zonula occludens
(C) macula adherens
(D) bile canaliculi
(E) hepatic sinusoids

370. The structures labeled with the arrows are

(A) cilia
(B) microvilli
(C) extruded cell cytoplasm
(D) hemosiderin
(E) endosomes

Questions 371–374

The following four questions refer to the photomicrograph below of a plastic-embedded, thin section.

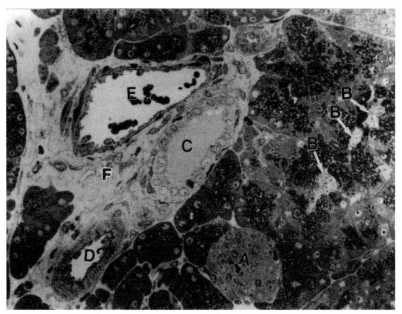

Courtesy of Eileen Roach.

371. Identify the organ.

(A) Submandibular gland
(B) Parotid gland
(C) Sublingual gland
(D) Pancreas
(E) Liver

372. The structure labeled A is

(A) a parasympathetic ganglion
(B) a central vein
(C) a hepatic portal vein
(D) an islet of Langerhans
(E) a bundle of nerve fibers in cross-section

373. The structures labeled B are

(A) centroacinar cells
(B) hepatocytes
(C) acinar cells
(D) intralobular duct cells
(E) Kupffer cells

374. The structure labeled C is

(A) an artery
(B) a hepatic portal vein
(C) a nerve
(D) a lymphatic
(E) a duct

375. The accompanying photomicrograph illustrates which of the following organs?

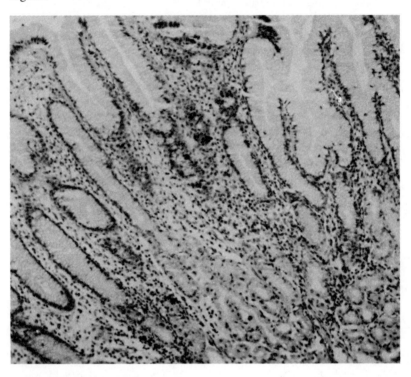

(A) Fundus
(B) Pylorus
(C) Colon
(D) Small intestine
(E) Esophagus

376. The photomicrograph below is of the junction between the

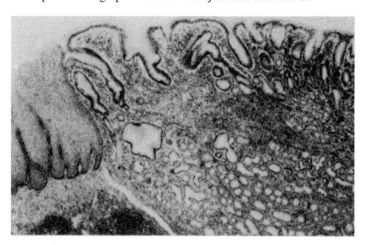

(A) anal canal and rectum
(B) esophagus and stomach
(C) skin of the face and mucous epithelium of the lip
(D) stomach and duodenum
(E) vagina and cervix

377. A unique histologic feature helpful in identifying the tissue shown in the photomicrograph below is the presence of

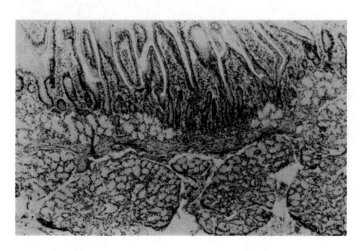

(A) Brunner's glands
(B) chief cells
(C) parietal cells
(D) Peyer's patches
(E) serous cells

378. The accompanying photomicrograph is of an organ that

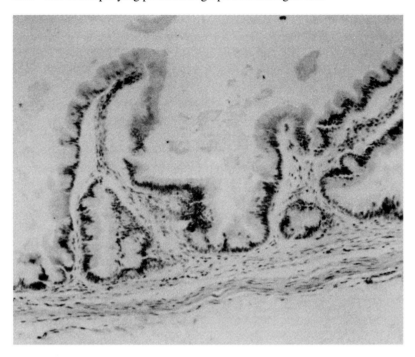

(A) stores and concentrates the bile
(B) synthesizes bile
(C) reabsorbs water
(D) exchanges nutrients and waste products between mother and child
(E) absorbs nutrients

379. The cells labeled with the asterisks in the electron micrograph are

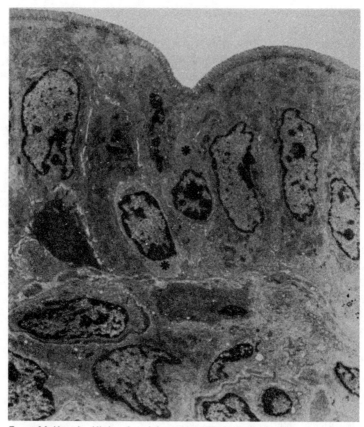

From McKenzie, Klein, *Am J Anat* 164:175–186, 1982, with permission.

(A) enterocytes
(B) goblet cells
(C) Paneth cells
(D) intraepithelial lymphocytes
(E) enteroendocrine cells

380. In the liver lobule shown in the photomicrograph below,

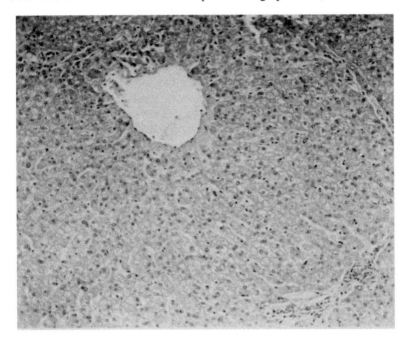

(A) bile and blood flow from the periphery to the center
(B) bile flows from the periphery to the center and blood flows from the center to the periphery
(C) bile flows from the center to the periphery and blood flows from the periphery to the center
(D) bile and blood flow from the center to the periphery
(E) bile is not produced

381. During the organogenesis of the gut,

(A) endoderm can independently differentiate into adult-type epithelium in the absence of mesenchyme

(B) mesenchyme differentiates into connective tissue and muscle in the absence of epithelium

(C) undifferentiated intestinal crypt cells have the competence to form the four basic cell types of the small intestinal epithelium

(D) hormones such as glucocorticoids inhibit epithelial differentiation through receptors on the epithelium

(E) extracellular matrix is not required for epithelial differentiation

382. Stem cells for organs of the gastrointestinal tract

(A) are found in the superficial layers of the epithelium of the esophagus

(B) are found in the base of the fundic glands of the stomach

(C) are found in the upper portion of the crypts in the large intestine

(D) are found in the villus compartment of the small intestine

(E) are the source of the four major cell types of the small intestinal epithelium

DIRECTIONS: Each numbered question or incomplete statement below is NEGATIVELY phrased. Select the **one best** lettered response.

Questions 383–384

A somewhat obese, 42-year-old mother of three children has experienced several episodes of severe pain in the upper right abdominal quadrant accompanied by pale-colored stools. Though she is not currently experiencing pain, the examiner notes that her skin and sclerae are somewhat yellow. A blood test indicates elevated bilirubin conjugated to glucuronic acid.

383. Elevated bilirubin levels in the blood can result from all the following EXCEPT

(A) neonatal deficiency of a hepatic enzyme that makes bilirubin soluble
(B) hepatocellular damage
(C) increased destruction of red blood cells
(D) obstruction of the common bile duct
(E) obstruction of the cystic duct

384. Hyperbilirubinemia may be caused by a number of different defects in hepatocyte metabolism, including all the following EXCEPT

(A) decreased uptake of bilirubin derived from the breakdown of hemoglobin
(B) inability of the hepatocyte to conjugate bilirubin
(C) deficiency in glucuronyl transferase
(D) increased transfer of bilirubin glucuronide into the bile canaliculi
(E) decreased conjugation of bilirubin with glucuronide in the smooth endoplasmic reticulum

385. All the following are functions of epithelial cells lining the small intestinal epithelium EXCEPT

(A) secretion of mucus
(B) synthesis of IgA
(C) absorption of digested nutrients
(D) completion of digestion initiated in the lumen
(E) synthesis of gastrointestinal hormones

386. Functions of hepatocytes include all the following EXCEPT

(A) synthesis of proteins including albumin and fibrinogen
(B) bile secretion
(C) storage of vitamin A
(D) detoxification of drugs such as barbiturates
(E) breakdown of hemoglobin

387. All the following are components of saliva present in the oral cavity EXCEPT

(A) amylase
(B) isotonic sodium concentration
(C) mucins
(D) lactoferrin
(E) IgA-secretory component complex

DIRECTIONS: Each group of questions below consists of lettered headings followed by a set of numbered items. For each numbered item select the **one** lettered heading with which it is **most** closely associated. Each lettered heading may be used **once, more than once, or not at all.**

Questions 388–391

Match each numbered description with the appropriate label on the diagram below of a coronal section of a human tooth

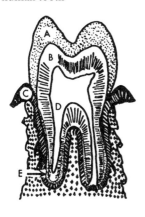

388. Has a composition similar to that of bone and is produced by cells similar in appearance to osteocytes

389. Is formed on a collagenous matrix that is resorbed upon mineralization by the same cells that secreted it

390. Contains abundant nerves, vessels, and loose connective tissue and odontoblasts

391. Consists of mineralized collagen secreted by cells derived from the neural crest

Questions 392–394

The diagram below shows the relationship between the esophagus, stomach, and duodenum. Match each description with the appropriate lettered site.

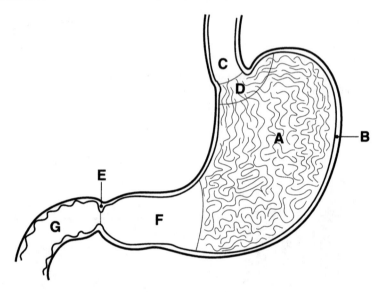

392. The region in which the majority of parietal cells are found

393. The region in which Brunner's glands are found

394. The site of glands with deep pits that produce mucus and a hormone that stimulates parietal cell HCl production

Questions 395–398

Match each numbered item with the correct lettered response.

(A) Activation of pancreatic zymogens

(B) Major secretagogue for enzymatic secretion in the pancreatic juice

(C) Major stimulus for salivary gland secretion

(D) Secretagogue for pancreatic ductal secretions

(E) ADP-ribosylation of G_s of the GTP-binding protein

(F) Decreased cyclic AMP in enterocytes

395. Secretin

396. Cholecystokinin

397. Enterokinase

398. Secretory diarrhea from cholera toxin

Questions 399–403

Match each function with the appropriate cell type.

(A) Paneth cells

(B) Parietal cells

(C) Enteroendocrine cells

(D) Chief cells

(E) Enterocytes

(F) Goblet cells

(G) M cells

(H) Mucous neck cells

399. Secretion of pepsinogen

400. Secretion of gastrin

401. Regulation of intestinal flora

402. Sampling of antigens from the intestinal lumen

403. Secretion of gastric intrinsic factor

Gastrointestinal Tract and Glands

Answers

358. The answer is D. *(Johnson, 3/e, pp 1123–1124. Junqueira, 8/e, pp 282, 284–285. Ross, 3/e, pp 448–449. Rubin, Lab Invest 19:598–626, 1968. Stevens, pp 161–162.)* In the resting parietal cell, the proton pump (H+,K+-ATPase) is found in the tubulovesicle membranes, which are located intracellularly. Upon activation of the parietal cell through calcium and diacylglycerol second messengers, the tubulovesicle membranes fuse with the plasma membrane by exocytosis. Histamine along with gastrin and acetylcholine activates the parietal cell. In the activated parietal cell, the Na+,K+-ATPase and the chloride channel of the apical plasma membrane maintain the appropriate ionic gradients to facilitate acid secretion. Carbonic anhydrase, a cytoplasmic enzyme, catalyzes the formation of carbonic acid (H_2CO_3) from carbon dioxide, which is the source of protons in the parietal cell and other cell types, such as the osteoclast, that also depend upon a proton pump. After dissipation of the stimulus (i.e., gastrin, acetylcholine, or histamine) or blockage with an antiulcerative drug (e.g., an H_2-blocker), the parietal cell returns to the resting state, which involves the endocytosis of membrane reforming the tubulovesicular arrangement within the cytoplasm. The sequestration of the proton pump and carbonic anhydrase in intracellular tubulovesicles in the resting state prohibits secretion.

359. The answer is D. *(Isselbacher, 13/e, pp 226–228. Junqueira, 8/e, pp 315–317, 320–322.)* Bile consists of bilirubin and bile acids. Bilirubin is derived from the breakdown of hemoglobin from senescent erythrocytes and ineffective erythropoiesis as well the breakdown of other heme-containing molecules. The breakdown process is carried out by phagocytic cells such as Kupffer cells in the liver. Approximately 90 percent of bile acids are recirculated through the intestine back to the hepatocyte, while 10 percent are synthesized by the hepatocyte.

360. The answer is E. *(Ito, Gastroenterology 88:250–260, 1985.)* Substances such as alcohol damage the epithelium of the stomach despite the presence of extensive mucus. Surface damage in the stomach is repaired by movement of lamellipodia of uninjured cells to cover the bare patches of basal lamina. This process is known as *restitution*. It occurs rapidly and does not require the migration of new cells from the neck or pit region of the fundic glands; however, at a later time new cells will proliferate to repopulate the upper layers of the gastric glands. Stem cells for the gastric glands are found in the neck and pit regions of the glands. Crypts are found in the small and large intestine, not in the stomach.

361. The answer is E. *(Junqueira, 8/e, pp 72–73, 292–293. Ross, 3/e, pp 389, 436–437. Stevens, pp 37, 162, 168.)* The enteroendocrine cells are unicellular endocrine glands much as the goblet cells are unicellular mucous glands. Enteroendocrine cells contain electron-dense granules, but have been best identified with antibodies to specific peptides and other enteroendocrine-derived secretions. They secrete into the bloodstream (endocrine function) or into the local area to affect nearby cells (paracrine function). The enteroendocrine cells may be identified by their staining response to silver or chromium stains—hence the older terms argentaffin and enterochromaffin, respectively. Examination of such preparations indicates that the enteroendocrine cells are rare compared with other mucosal cell types, including the mucous cells. Holocrine secretion occurs in the sebaceous glands and involves the destruction of the cell in association with secretion. Apocrine secretion occurs in the sweat glands and involves the loss of the apical portion of the cell in conjunction with secretion.

362. The answer is B. *(Junqueira, 8/e, pp 292–293. Ross, 3/e, pp 277–280, 451, 458–459. Stevens, pp 159, 286–287.)* The enteroendocrine cells and the enteric (intrinsic) nervous system secrete similar peptides and are found throughout the gastrointestinal tract. Although there has been some controversy regarding the origin of enteroendocrine cells, it is now well documented that these cells are derived from the same stem cell as other epithelial cell types and originate embryonically from the endoderm. These cells turn over at a slower rate than other epithelial cell types. In contrast, the cells that compose the enteric nervous system are neurons, derived from neural crest. There is little cell replacement except in the glial populations. The enteric nervous system, particularly the myenteric (or Auerbach's) plexus, is responsible for the intrinsic rhythmicity of the gut. The enteroendocrine cells function in local paracrine regulation of the mucosa (e.g., acid secretion in the stomach, mucosal growth, small intestinal secretion, and turnover).

363. The answer is B. *(Junqueira, 8/e, p 295. Ross, 3/e, p 456. Stevens, p 261.)* Lipid digestion occurs in the small intestinal lumen by the action of bile (from the liver and bile duct) and lipase (from the pancreas). Bile serves to emulsify the lipid to form micelles, while lipase breaks down the lipid from triglycerides to fatty acids, glycerol, and monoglycerides. These three breakdown products diffuse freely across the microvillus border to enter the apical portion of the enterocyte. Triglycerides are resynthesized in the smooth endoplasmic reticulum. Proteins are synthesized in the RER and are combined with sugar and lipid portions in the Golgi to form glycoproteins and lipoproteins. These two types of molecules form the coverings of the triglyceride cores of the chylomicra. The chylomicra are released at the basolateral membranes by exocytosis into the lacteals. From the lacteals the chylomicra travel into the cisterna chyli and eventually into the venous system by way of the thoracic duct. Digestion of fat occurs to a greater extent in the duodenum and jejunum than in the ileum.

364. The answer is D. *(Isselbacher, 13/e, pp 220, 296, 899–900, 1420. Junqueira, 8/e, p 270. Stevens, pp 172–173.)* Hirschsprung disease and Chagas disease have different etiologies, but both inhibit intestinal motility by affecting the myenteric (Auerbach) plexus. The submucosal (Meissner) plexus is more involved in regulation of lumenal size and therefore will affect defecation, but will be less involved in peristalsis. Vascular smooth muscle, the muscularis mucosa, and enteroendocrine cells do not play a major role in the regulation of peristalsis, which is observed even after removal of the gut and placement in a nutrient solution. Hirschsprung disease, also known as *aganglionic megacolon,* results from failure of normal migration of neural crest cells to the colon, resulting in an aganglionic segment. Although both the myenteric and submucosal plexuses are affected, the primary regulator of intrinsic gut rhythmicity is the myenteric plexus. Chagas disease is caused by the protozoan *Trypanosoma cruzi.* Severe infection results in extensive damage to the myenteric neurons.

365. The answer is D. *(Guyton, 8/e, pp 711–712.)* The primary secretion produced by the acinar cells consists of amylase, mucus, and ions in the same concentrations as those of the extracellular fluid. In the duct system, Na^+ is actively absorbed from the lumen of the ducts, Cl^- is passive absorbed, K^+ is actively secreted, and HCO_3^- is secreted. The result is a hypotonic sodium and chloride concentration and a hypertonic potassium concentration.

366. The answer is B. *(Guyton, 8/e, pp 712–713.)* The parasympathetic nervous system carries signals that originate in the salivatory nuclei of the medulla and pons. The parasympathetic input is the primary neural regulator of the gland, although the sympathetic nervous system also influences secretion. Elevated aldosterone levels affect the amount and ionic concentration of the saliva, resulting in decreased NaCl secretion and increased K^+ concentration. Cholecystokinin (pancreozymin) and secretin are the hormones that regulate acinar and ductal secretions, respectively, in the exocrine pancreas.

367. The answer is E. *(Johnson, 3/e, pp 1123–1126. Junqueira, 8/e, pp 282–285.)* The parietal cell is maintained by a series of pumps and exchanger systems. The Na⁺, K⁺-ATPase pump on the basolateral membranes is responsible for the extrusion of sodium and the accumulation of potassium. The gastric proton pump transports protons into the lumen in exchange for potassium. The proton pump is located in the apical membrane along with the chloride channel. Carbonic anhydrase, an enzyme present apically as well as in the cytoplasm, is available for the formation of bicarbonate and protons when the parietal cell is stimulated. In the resting state carbonic anhydrase and the proton pump are located in the tubulovesicular membranes and do not have access to the apical surface. Exocytosis of tubulovesicles results in the availability of the proton pump and chloride channels. HCO_3^- is released across the basal membrane, which results in an increase in blood pH during digestion.

368. The answer is B. *(Burkitt, 3/e, pp 18–19. Junqueira, 8/e, pp 45–46, 312–314, 316. Ross, 3/e, pp 42–43.)* Cells have several different types of cytoplasmic inclusions, including lipid droplets and glycogen. The hepatocyte, under the regulation of insulin and glucagon, stores glucose in its polymerized form of glycogen. In electron micrographs, glycogen appears as scattered dark particles with an approximate diameter of 15 to 25 nm. Lipid droplets appear as spherical, homogeneous structures of varying density and diameter, although their diameter would be considerably larger than that of the glycogen granules. Ribosomes are found on the rough endoplasmic reticulum or as free structures, in which case they are not found in clusters like glycogen.

369–370. The answers are 369-D, 370-B. *(Junqueira, 8/e, pp 312–314, 316. Ross, 3/e, pp 502–505, 524–525.)* The bile canaliculi are formed as the space between the lateral surfaces of adjacent hepatocytes. Microvilli line the bile canaliculi and are visible protruding into the lumen. The membranes between the cells are connected by tight (zonula occludens) and gap (nexus) junctions, neither of which are visible in the photomicrograph.

371–374. The answers are 371-D, 372-D, 373-A, 374-E. *(Burkitt, 3/e, pp 279–281. Junqueira, 8/e, pp 303–308. Ross, 3/e, pp 528–529.)* The organ in the photomicrograph is the pancreas. The pancreas functions as both an exocrine (secretion of pancreatic juice) and endocrine (secretion of insulin and glucagon) gland. The islets (A) have a heterogeneous distribution within the pancreas (i.e., they decrease from the tail to the head of the gland) and may be used to distinguish the pancreas from the parotid gland. The submandibular and sublingual glands can be ruled out because of the purely serous nature of the acini within the exocrine portion of the gland. The centroacinar cells (B) are modified intralobular duct cells, specifically from the intercalated duct, and are present in the lumen of each acinus. The duct (C) can be distinguished by the presence of a cuboidal epithelium, the absence of blood and blood cells from the lumen, and the absence of a characteristic vascular wall. A pancreatic artery (D) and a vein (E) are shown within the interlobular connective tissue (F).

375. The answer is B. *(Burkitt, 3/e, p 257. Erlandsen, pp 113, 115. Junqueira, 8/e, p 277. Ross, 3/e, pp 446–450. Stevens, p 164.)* The photomicrograph shows the pylorus of the stomach. The pylorus differs from the fundus of the stomach in the length of the pits of the glands compared to the length of the gland. In the fundus there are short pits and long glands (pit:gland ratio of about 1:4) compared with a pit:gland ratio of 1:2 in the pylorus. There is also an absence of parietal cells in the pylorus.

376. The answer is B. *(Junqueira, 8/e, pp 276–278, 298, 433, 440. Ross, 3/e, pp 474–477, 726–727, 732–733. Stevens, pp 155, 174–175, 325, 327.)* The histologic section in the photomicrograph accompanying the question shows two distinctly different types of epithelium. The esophageal mucosa on the left is nonkeratinized, stratified squamous epithelium overlying a fibrovascular submucosa and smooth muscle. The gastric mucosa on the right is simple columnar epithelium with simple glands, overlying submucosa and smooth muscle. Skin is keratinized. Both the stomach and duodenum are simple columnar epithelium. The simple columnar epithelium of the distal-most rectum contains only mucous cells. The cervical mucosa contains extensive cervical glands and the vaginal epithelium is keratinized.

377. The answer is A. *(Burkitt, 3/e, pp 258–260. Erlandsen, p 110. Junqueira, 8/e, p 293. Ross, 3/e, pp 488–489, 526–527. Stevens, pp 99–100, 169.)* The presence of Brunner's glands in the submucosal layer of the intestine is an identifying feature of the duodenum. Parietal cells are unique to the stomach (with the exception of Meckel's diverticulum), whereas Peyer's patches of lymphoid tissue are most characteristic of the ileum. Chief cells are found in the fundus and body of the stomach. Serous cell are found in numerous locations throughout the alimentary tract.

378. The answer is A. *(Burkitt, 3/e, p 278. Erlandsen, pp 124–129. Junqueira, 8/e, pp 320–322. Ross, 3/e, pp 507–509, 526–527. Stevens, pp 184–185.)* The photomicrograph illustrates the structure of the gallbladder.

Although the fingerlike extensions resemble villi, they represent changes that occur in the mucosa with increasing age. The thinness of the wall is the notable characteristic of the gallbladder. This organ is responsible for the storage and concentration of the bile, which is synthesized by hepatocytes and transported to the gallbladder.

379. The answer is D. *(Ross, 3/e, pp 459–461, 470–471. Stevens, p 169.)* Intraepithelial lymphocytes are lymphocytes that have crossed the basal lamina. These cells may respond to antigen in the lumen of the small intestine or antigen that has been sampled from the lumen and processed by M cells in the Peyer's patches. Enterocytes are the absorptive cells of the gut and possess numerous microvilli on their apical surfaces. Goblet cells are filled with mucus. Paneth cells and enteroendocrine cells contain granules but differ in function.

380. The answer is C. *(Burkitt, 3/e, pp 271–273. Erlandsen, pp 126–127. Junqueira, 8/e, pp 306–312, 314, 316. Ross, 3/e, pp 496–502. Stevens, pp 176, 180–181.)* In a liver lobule, bile and blood flows occur in opposite directions. Blood from the hepatic artery and hepatic portal vein flows through the sinusoids. The direction of blood flow is from the portal triad (hepatic artery, hepatic portal vein, and bile duct) toward the central vein (in the center of the lobule). The direction of bile flow is the reverse of blood flow. Bile is formed by the hepatocytes and is released into bile canaliculi, which are located between the lateral surfaces of adjacent hepatocytes. The direction of flow is from the hepatocytes toward the bile duct, which drains bile from the liver on its path to the gallbladder, where the bile is stored and concentrated.

The lobule concept of the liver is difficult to visualize in human liver because of the irregularity of morphological components. The classic lobule emphasizes endocrine function in the liver (from the periphery to the center). The portal lobule is formed by three central veins surrounding a central portal triad and emphasizes the exocrine function of the liver, i.e., bile flow. The liver acinus is based upon real blood flow since neither the hepatic portal veins nor the hepatic arteries terminate at the portal triad but create a three-dimensional pattern as they form smaller and smaller vessels. The liver acinus is a diamond-shaped structure based upon blood supply around the smaller branches of these vessels. In the photomicrograph, the central vein is located near the top left and a portal triad is found near the bottom right. Blood flows from the triad toward the central vein, while bile flows toward the triad.

381. The answer is C. *(Lebenthal, pp 19–35.)* During the development of the gut, endoderm requires the presence of mesenchyme and mesenchyme requires the presence of epithelium. Hormones such as glucocorticoids stimulate precocious development of the gut through glucocorticoid receptors found on mesenchymal cells, not epithelial cells. The effect of the glucocorticoids is mediated by a mesenchyme-derived factor, which stimulates the differentiation of the epithelial cells. Extracellular matrix components are required for the normal differentiation of the endoderm to form the gut epithelium. Undifferentiated intestinal crypt cells, such as those derived from an intestinal carcinoma, can act as stem cells to reconstitute the entire epithelium. The undifferentiated cells differentiate into the four basic cell types of the intestinal epithelium: goblet, enteroendocrine, Paneth, and intestinal absorptive (enterocytes).

382. The answer is E. *(Junqueira, 8/e, pp 299–300. Ross, 3/e, pp 451–452, 464, 466. Stevens, pp 162, 168, 169.)* Cell renewal in the gastrointestinal tract is more rapid than in other continuously renewing epithelia such as the epidermis. These renewing cells of the gastrointestinal tract are therefore very sensitive to antimitotic drugs. Stem cells are found in each organ in the gastrointestinal tract. In the esophagus and anus the stem cells are found in the proliferative zone located in the basal layer of the stratified squamous epithelium. In the fundus, the proliferative zone is located in the neck and pit regions of the glands. The cell types of the fundic glands arise from these stem cells. Surface mucous and mucous neck cells move upward fron the pit while parietal and chief cells move downward from their site of origin. In the large intestine, the proliferative zone is located in the basal region of the crypts and enterocytes and goblet cells move upward as they differentiate and move toward the surface. In the small intestine, all four epithelial cell types are derived from a single stem cell. Paneth cells move downward from their site of origin, while goblet, enteroendocrine, and absorptive cells move upward on the villus as they differentiate.

383. The answer is E. *(Junqueira, 8/e, pp 315–316, 320–322. Ross, 3/e, pp 497, 506–507. Stevens, p 182.)* Bilirubin, a product of iron-free heme, is liberated during the destruction of old erythrocytes by the mononuclear macrophages of the spleen and, to a lesser extent, of the liver and bone marrow. The hepatic portal system brings splenic bilirubin to the liver, where it is made soluble for excretion by conjugation with glucuronic acid. Commonly, initial low levels of glucuronyl transferase in the underdeveloped smooth endoplasmic reticulum of hepatocytes in the newborn result in jaundice (neonatal hyperbilirubinemia); less commonly, this enzyme is genetically lacking. The ability of mature hepatocytes to take up and conjugate bilirubin may be exceeded by abnormal increases in erythrocyte destruction (hemolytic jaundice) or by hepatocellular damage (functional jaundice), such as in hepatitis. Finally, obstruction of the duct system between the liver and duodenum (usually of the common bile duct in the adult and rarely from aplasia of the duct system in infants) results in a backup of bilirubin (obstructive jaundice). However, obstruction of the cystic duct, while painful, will not interfere with the flow of bile from the liver to the duodenum.

384. The answer is D. *(Isselbacher, 13/e, pp 1453–1456. Junqueira, 8/e, pp 320–322.)* Increased plasma levels of bilirubin (hyperbilirubinemia) result from increased bilirubin turnover, impaired uptake of bilirubin, or decreased conjugation of bilirubin. Increased bilirubin turnover occurs in Dubin-Johnson and Rotor syndromes, in which there is impairment of the transfer and excretion of bilirubin glucuronide into the bile canaliculi. In Gilbert syndrome, there is impaired uptake of bilirubin into the hepatocyte and a defect in glucuronyl transferase. In Crigler-Najjar syndrome, a defect in glucuronyl (glucuronosyl) transferase occurs in the neonate.

385. The answer is B. *(Alberts, 3/e, pp 1210–1211. Junqueira, 8/e, pp 269–270. Ross, 3/e, pp 453–459. Stevens, pp 165–170.)* The small intestinal lining cells include goblet cells, which secrete mucus. They also include enterocytes (intestinal absorptive cells), which absorb digested nutrients and are covered at their apical surface by microvilli with brush border enzymes that complete digestion begun in the small intestinal lumen. Enteroendocrine cells synthesize numerous hormones, which may be released into the bloodstream or between adjacent cells. The small intestinal epithelium also synthesizes secretory component, which is added to IgA secreted by plasma cells in the lamina propria. The IgA undergoes transcytosis through a receptor-mediated process across the enterocytes, at which time the secretory component is added to the IgA. The secretory component protects the IgA, which is released into the small intestinal lumen and carries out the functions of secretory immunity.

386. The answer is E. *(Junqueira, 8/e, pp 314–318. Ross, 3/e, pp 502–506. Stevens, pp 178–180.)* Hepatocytes are involved in the many functions of the liver including protein synthesis, bile secretion, storage of metabolites (e.g., vitamin A), and detoxification of drugs. Bile consists of bilirubin derived from red blood cell breakdown and bile acids, which are primarily recycled from the small intestine and partially synthesized by the hepatocytes. Detoxification occurs by both oxidation and conjugation to sulfate or glucuronide. During detoxification the smooth endoplasmic reticulum is vastly enlarged in response to toxins or drugs. The breakdown of hemoglobin is important for the recycling of bilirubin, but it is carried out by monocyte-derived cells in the body including the Kupffer cells, which are the antigen-presenting, phagocytic cells of the liver. While the Kupffer cells are a self-limited population, they are derived from monocytes at some time during development. In addition, hemoglobin breakdown occurs in other organs, particularly the spleen.

387. The answer is B. *(Guyton, 8/e, p 711. Junqueira, 8/e, pp 269–270, 301–303.)* The saliva contains water, enzymes (amylase and lysozyme), mucins, lactoferrin, and secretory IgA. Amylase digests starches; mucins primarily lubricate; lysozyme hydrolyzes bacterial cell walls; lactoferrin is an iron-binding protein; and IgA is a secretory immunoglobulin involved in the host defenses of the oral cavity. Saliva contains less sodium than plasma and therefore is hypotonic.

388–391. The answers are 388-E, 389-A, 390-D, 391-B. *(Junqueira, 8/e, pp 273–280. Ross, 3/e, pp 409–415. Stevens, pp 146–152.)* The pulp of a mature tooth (labeled D in the diagram) consists primarily of loose connective tissue rich in vessels and nerves. Odontoblasts, which are derived from the neural crest, lie at the edge of the pulp cavity and secrete collagen and other molecules, which mineralize to become dentin (B). Mineralization of the matrix occurs around the process of odontoblasts and forms dentinal tubules. Ameloblasts, which are ectodermal derivatives, lay down an organic matrix and secrete enamel, initially onto the surface of the dentin. As hydroxyapatite crystals form at the apices of ameloblast (Tomes') processes, rods of enamel grow peripherally and the ameloblasts resorb the organic matrix so that the enamel layer (A) is almost entirely mineral. Upon eruption of the tooth, enamel deposition is complete and the ameloblasts are shed. Cementum (E) has a composition similar to that of bone, is produced by cells similar in appearance to osteocytes, and covers the dentin of the root. The periodontal ligament (C) consists of coarse collagenous fibers running between the alveolar bone and the cementum of the tooth and separates the tooth from the alveolar socket. Although the periodontal ligament suspends and supports each tooth, the membrane permits physiologic movement within the limits provided by the elasticity of the tissue.

392–394. The answers are 392-A, 393-G, 394-F. *(Burkitt, 3/e, p 252. Junqueira, 8/e, pp 277, 280–289, 293)* The diagram shows the anatomical relationship between the esophagus, stomach, and duodenum. The esophagus (C) joins the stomach in the cardiac region (D). The mucosa of the fundus and body (A) is composed of gastric glands containing mucous cells, chief cells that synthesize pepsinogen, and parietal cells that synthesize hydrochloric acid and gastric intrinsic factor. The pylorus (F) contains glands with deeper pits than those of the fundus and body. These glands contain more mucous cells and many gastrin-secreting enteroendocrine cells. Food entering the pylorus stimulates the release of gastrin, which stimulates HCl production by the parietal cells. The pylorus connects with the duodenum (G), which contains the mucus and bicarbonate-neutralizing secretion of the Brunner's glands. The wall of the stomach consists of the mucosa (epithelium, lamina propria, and muscularis mucosa), submucosa, muscularis externa, and serosa (B) lined by a mesothelium.

395–398. The answers are 395-D, 396-B, 397-A, 398-E. *(Donowitz, pp 241–262. Junqueira, 8/e, pp 312–320, 329–330, 333. Ross, 3/e, pp 415–422, 463–464, 507–509. Stevens, p 181.)* The exocrine pancreas and the salivary glands have a similar acinar-ductal structure, although there are some variations. For example, the pancreatic acini lack myoepithelial cells but contain centroacinar cells. Myoepithelial cells surround the acini and portions of the smaller (intralobular: intercalated and striated) ducts and stimulate the secretion of amylase and other salivary components into the oral cavity. The centroacinar cells, which are found only in the pancreas, are the first portion of the duct system and are found in the lumina of the pancreatic acini.

Pancreatic secretion is regulated by hormones. Secretin regulates ductal secretion while cholecystokinin regulates the release of enzymes (amylase, lipase, DNAse, RNAse, and the other enzymes that compose the pancreatic juice). A number of pancreatic secretions are released into the pancreatic duct system as zymogens (inactive precursors). They are activated only when they arrive in the small intestinal lumen. Enterokinase, a brush border enterocyte enzyme, converts trypsinogen to trypsin. Trypsin and enterokinase are responsible for the activation of chymotrypsinogen, proelastase, and procarboxypeptidase A and B to their active forms: chymotrypsin, elastase, and carboxypeptidase A and B. By contrast, the primary regulator of secretion from the salivary glands is the parasympathetic nervous system. Activation begins with enterokinase, which is an enzyme produced by the small intestinal mucosa.

The gallbladder does not synthesize bile but serves to store and concentrate the bile. In response to cholecystokinin the gallbladder contracts and releases bile into the small intestinal lumen by way of the bile duct. Bile is important in the emulsification of lipids, which facilitates digestion of fat in the small intestine.

Cholera toxin causes secretory diarrhea through the ADP-ribosylation of G_s of the GTP-binding protein, which leads to elevated cyclic AMP and the opening of the chloride channel. The exit of chloride through the open channels is followed by the passage of sodium and water. The result can be dehydration, which can be offset by intraveous feeding or oral rehydration therapy.

399–403. The answers are 399-D, 400-C, 401-A, 402-G, 403-B. *(Junqueira, 8/e, pp 284–287, 290–294. Ross, 3/e, pp 446–447, 455–459. Stevens,* *pp 160–169.*) The cell types listed in this question group are found in the fundic and pyloric glands of the stomach, the crypts of Lieberkühn of the small intestine, and the follicle-associated epithelium of the Peyer's patches. The fundic glands contain surface mucous, mucous neck, parietal, and chief cells. The surface mucous and mucous neck cells produce mucus that protects the surface of the stomach from the highly acidic milieu. The parietal cells produce acid and gastric intrinsic factor. The latter is formed in the stomach and binds to vitamin B_{12}, which is required for its uptake from the ileum. The chief cells produce the proenzyme pepsinogen, which is released into the lumen of the stomach where the acidic pH converts it to its active form (pepsin).

The crypts of Lieberkühn contain the following cell types: goblet, Paneth, enteroendocrine, and absorptive cells. The absorptive cells, also known as *enterocytes,* are responsible for the absorption of metabolites from the small intestinal lumen. These cells are intimately involved in the breakdown of fats, proteins, and carbohydrates. Digestion is facilitated by pancreatic juice, which arrives by way of the pancreatic duct; bile, which is stored in the gallbladder; and the brush border enzymes, which are located on the apical surface of the enterocytes. The goblet cells are named for their shape and are responsible for the production of mucus in the small intestine. The enteroendocrine cells are found throughout the gastrointestinal tract. In the pylorus, the enteroendocrine cells produce gastrin. In the remainder of the stomach, enteroendocrine cells synthesize glucagon. In the small intestine, enteroendocrine cells synthesize secretin, cholecystokinin, glicentin (glucagon-like substance, or enteroglucagon), somatostatin, vasoactive intestinal polypeptide, serotonin, substance P, motilin, and gastric inhibitory peptide. Paneth cells are found at the base of the crypts and are responsible for the production of lysozyme and the selective phagocytosis of bacteria and protozoans. They are therefore involved in the regulation of the flora of the small intestine. Paneth cells are absent in diseases of diminished nutritional status, such as kwashiorkor in children. When a normal nutritional status is reestablished, the Paneth cells return to normal numbers.

The M cells are found in the follicle-associated epithelium of the Peyer's patches and sample antigens from the lumen. Endocytosis of luminal antigens by M cells is followed by presentation of antigen to intraepithelial lymphocytes. The Peyer's patches form part of the gut-associated lymphoid tissue (GALT). Other components include lymphocytes of the lamina propria and the epithelium, which may be both B and T cells. Unlike transient lymphocytic infiltrates, the Peyer's patches are permanent structures with B- and T-cell–specific areas and high endothelial postcapillary venules with their surface addressins, which are important in mucosal targeting of lymphocytes with specific homing receptors.

Endocrine Glands

DIRECTIONS: Each question below contains five suggested responses. Select the **one best** response to each question.

404. The adrenal cortex influences the secretion of the adrenal medulla by

(A) secretion of aldosterone into the intraadrenal circulation
(B) secretion of glucocorticoids into the intraadrenal circulation
(C) autonomic neural connections
(D) secretion of monoamine oxidase into the portal circulation
(E) secretion of androgens into the intrarenal circulation

405. In contrast to the adrenal medulla, adrenocortical cells

(A) release their secretions by exocytosis
(B) are derived embryologically from the neural crest
(C) do not store appreciable quantities of hormones
(D) are independent of anterior pituitary regulation
(E) receive blood supply from all three suprarenal arteries

406. The function of the fetal (provisional) adrenal cortex is to produce

(A) estradiol
(B) glucocorticoids
(C) aldosterone
(D) androgens
(E) ACTH

407. In Addison's syndrome, there is autoimmune destruction of the adrenal cortex. In this disorder, one would expect to observe hypertrophy of

(A) lactotrophs
(B) corticotrophs
(C) thyrotrophs
(D) spongiocytes
(E) gonadotrophs

408. A pheochromocytoma is a common tumor of the adrenal medulla. In the presence of this tumor, which of the following symptoms would most likely be observed?

(A) Hypotension
(B) Hyperglycemia
(C) Hirsutism
(D) Vasodilatation of arterioles
(E) A shift to increased production of norepinephrine

409. The pituitary resides in the _____ of the _____ bone

(A) squama, temporal
(B) sella turcica, sphenoid
(C) petrous portion, temporal
(D) greater wing, sphenoid
(E) base, occipital bone

410. In the β-cells of the pancreas, proinsulin is converted to insulin

(A) in the bloodstream
(B) in clathrin-coated vesicles
(C) in the Golgi
(D) at the time of translocation into the rough endoplasmic reticulum
(E) at the time of fusion of mature granules with the plasma membrane

411. Which of the following symptoms would one expect in the presence of an insulinoma?

(A) Hyperglycemia
(B) Anemia
(C) Decreased storage of glycogen in the liver
(D) Stimulation of hepatic phosphorylase
(E) Decreased glucose metabolism in muscle

412. The anterior lobe of the pituitary is not a good candidate for transplantation compared with other endocrine glands because

(A) more severe rejection of neurally related tissue occurs compared with other endocrine organs
(B) its hormonal source is unavailable after its axonal connections to the hypothalamus are disrupted
(C) it lacks function when separated from the hypothalamohypophyseal portal system
(D) neogenesis of blood vessels will not occur at the transplant site
(E) distribution of the superior hypophyseal arteries is unique

413. Hormonal secretion from the thyroid gland

(A) is independent of anterior pituitary regulation
(B) is stored as triiodothyronine (T_3) and thyroxine (T_4) in the follicular colloid
(C) occurs by exocytosis into the duct system
(D) includes a hormone that transiently increases blood calcium levels
(E) involves pinocytosis of iodinated thyroglobulin from the colloid

414. The gland shown in this photomicrograph

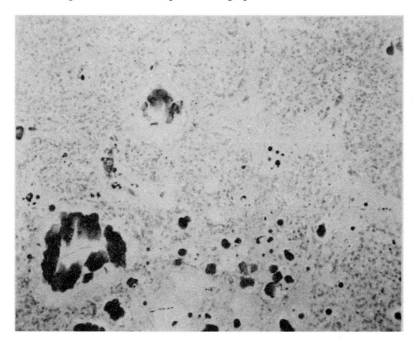

(A) arises as an outgrowth of the midbrain
(B) influences the rhymicity of other endocrine organs
(C) contains many melanocytes
(D) is innervated by preganglionic sympathetic fibers
(E) secretes melanocyte-stimulating hormone (MSH)

---

415. A female patient suffers from a deficiency in the enzyme involved in the last step of cortisol synthesis. Which of the following would most likely be associated with this condition?

(A) Masculinization of the female genitalia
(B) Decreased blood ACTH levels
(C) Atrophy of the zona reticularis
(D) Hypersecretion of cortisol
(E) Hypersecretion of vasopressin

416. Elevation of blood calcium

(A) is stimulated by parathyroid (PTH) secretion
(B) is accompanied by elevated blood PO_4^{2-} levels
(C) is independent of kidney function
(D) is independent of small intestinal function
(E) is independent of vitamin D levels

417. The gland shown in the light microscopic photograph below produces which of the following hormones?

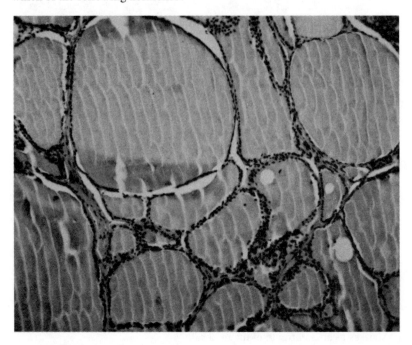

(A) Cortisone
(B) Milk
(C) Parathyroid hormone
(D) Melanocyte-stimulating hormone (MSH)
(E) Thyroxine

DIRECTIONS: Each numbered question or incomplete statement below is NEGATIVELY phrased. Select the **one best** lettered response.

418. All the following cells or parts of the pituitary are derived embryologically from Rathke's pouch (the oral ectoderm) EXCEPT

(A) gonadotrophs
(B) pars intermedia
(C) pars tuberalis
(D) pars nervosa
(E) lactotrophs

419. Surgical removal of the pituitary gland (hypophysectomy) results in a decrease in all the following EXCEPT

(A) activity of the thyroid follicular cells
(B) growth of ovarian follicles
(C) size of the adrenal zona fasciculata cells
(D) size of the parathyroid chief cells
(E) spermatogenesis

420. All the following are well-documented complications of diabetes mellitus EXCEPT

(A) retinopathy
(B) nephropathy
(C) liver fibrosis
(D) neuropathy
(E) cardiovascular disease

DIRECTIONS: Each group of questions below consists of lettered headings followed by a set of numbered items. For each numbered item select the **one** lettered heading with which it is **most** closely associated. Each lettered heading may be used **once, more than once, or not at all.**

Questions 421–422

Match each description with the correct structure labeled on the light microscopic photograph.

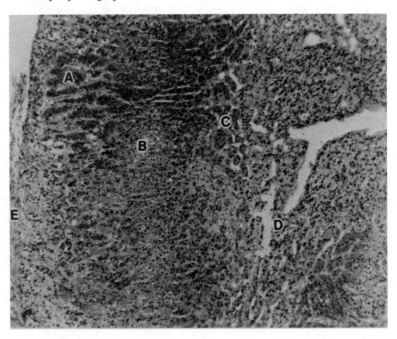

421. Regulation primarily by angiotensin II

422. Synthesis of androgens

Questions 423–424

Refer to the figure below to answer the following questions.

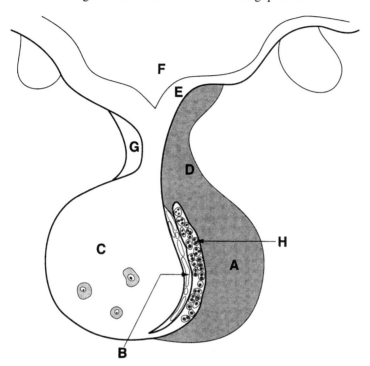

423. Which of the regions labeled on the figure is the site of action of corticotropin-releasing hormone (CRH)?

424. In which region would one expect to find storage of a secretory product that is synthesized in another organ?

Questions 425–427

Match each description with the hormone produced by the islets of Langerhans.

    (A) Insulin
    (B) Glucagon
    (C) Somatostatin
    (D) Vasoactive intestinal peptide (VIP)
    (E) Pancreatic polypeptide

425. Increases permeability of cell membranes to glucose

426. Regulates release of insulin and glucagon and is localized in δ-cells

427. Is normally secreted during hypoglycemic episodes

Endocrine Glands

Answers

404. The answer is B. *(Isselbacher, 13/e, p 413. Junqueira, 8/e, pp 387–388. Ross, 3/e, pp 609–610, 618–619.)* The adrenal gland functions as two separate glands. The adrenal cortex is derived from mesoderm and the adrenal medulla from neural crest. The blood supply to the adrenal is derived from three adrenal arteries: (1) the superior adrenal (suprarenal) from the inferior phrenic, (2) the middle adrenal from the aorta, and (3) the inferior adrenal from the renal artery. Most of the blood supply entering the medulla passes through the cortex. Glucocorticoids synthesized in the zona glomerulosa of the adrenal are released into the sinusoids and enter the medulla. Metabolism in the adrenal medulla is regulated by glucocorticoids since they induce the enzyme phenylethanolamine-*N*-methyltransferase, which catalyzes the methylation of norepinephrine to epinephrine. The adrenal gland is not usually considered a classic portal system although there are similarities. Monoamine oxidase is a mitochondrial enzyme that regulates the storage of catecholamines in peripheral sympathetic nerve endings.

405. The answer is C. *(Junqueira, 8/e, p 388.)* The cells of the adrenal cortex import cholesterol and acetate and produce different hormones with zone specificity (i.e., zona glomerulosa, aldosterone; zona fasciculata, glucocorticoids; zona reticularis, androgens). However, in all zones the cells do not store appreciable quantities of hormones, there is an absence of secretory granules, and the steroid hormones are released by diffusion through the plasma membrane without use of the exocytotic process used by most glands, including the adrenal medulla. The adrenal cortex is not derived from the neural crest; it forms from the mesoderm of the urogenital ridge. The adrenal medulla is independent of direct anterior pituitary regulation. The adrenal cortex is under the influence of ACTH synthesized by the anterior pituitary. Blood supply to both the cortex and medulla of the adrenal gland is derived from the superior, middle, and inferior suprarenal arteries.

406. The answer is D. *(Junqueira, 8/e, pp 392–393.)* The fetal adrenal cortex is part of the fetal-maternal-placental unit, which includes a number of different organs in the mother and fetus. It produces androgens that are converted by the placenta into estradiol. The fetal adrenal cortex involutes following birth, resulting in an overall reduction in the size of the adrenal. The adult cortex develops as the fetal adrenal degenerates.

407. The answer is B. *(Isselbacher, 13/e, pp 1970–1971. Junqueira, 8/e, pp 382, 384, 392, 394–395. Stevens, p 262.)* Loss of adrenocortical function would feed back to the anterior pituitary as well as the hypothalamus. The result would be elevated secretion of ACTH by corticotrophs and elevated secretion of ACTH-releasing factor from the hypothalamus. Spongiocytes are the cells that compose the zona fasciculata; they appear vacuolated or spongy because of the loss of lipid during routine histologic processing. The spongiocytes would be expected to undergo hypotrophy in adrenocortical insufficiency disorders such as Addison's syndrome. In Cushing's disease, there is hyperfunction of the adrenal cortex and hypertrophy of the spongiocytes. Other cells of the anterior pituitary include the lactotrophs, which synthesize prolactin; thyrotrophs, which synthesize thyroid-stimulating hormone (TSH); and gonadotrophs, which synthesize follicle-stimulating hormone (FSH) and luteinizing hormone (LH).

408. The answer is B. *(Isselbacher, 13/e, pp 1976–1977. Junqueira, 8/e, pp 387, 394.)* Pheochromocytoma is a common tumor of the adrenal medulla that leads to an excess of norepinephrine, which causes hypertension and hyperglycemia. Vasoconstriction of arterioles occurs in conjunction with the increased blood pressure.

409. The answer is B. *(Burkitt, 3/e, pp 304–306, 308. Ross, 3/e, pp 622–623. Stevens, pp 250–254.)* The pituitary, or hypophysis, is located in the sella turcica of the sphenoid bone. The middle ear bones are located in the petrous portion of the temporal bone.

410. The answer is B. *(Alberts, 3/e, pp 628–629.)* Insulin is synthesized as preproinsulin. The presequence is removed at the time of translocation into the rough endoplasmic reticulum. Proinsulin is converted to insulin in a cleavage process that occurs in clathrin-coated vesicles. The process involves acidification of the vesicles by an H^+ pump in the membrane.

411. The answer is C. *(Guyton, 8/e, p 857. Isselbacher, 13/e, pp 1540–1541.)* Tumors of the islets of Langerhans include insulinomas in which elevated levels of insulin are secreted into the bloodstream. The result is hypoglycemia as blood sugar levels drop. There is decreased storage of glycogen in the liver, inhibition of hepatic phosphorylase (which causes the breakdown of glycogen to form glucose), and increased glucose metabolism in muscle.

412. The answer is C. *(Junqueira, 8/e, pp 382–383. Ross, 3/e, pp 598–599. Stevens, pp 254–255.)* The anterior pituitary is unique in that it is dependent on the presence of the hypothalamohypophyseal portal system. Releasing and inhibitory factors are transported from the cell bodies in the hypothalamus along axons into the median eminence, where the secretion is released into a primary capillary plexus. The hypothalamohypophyseal portal system carries blood from the primary plexus to the secondary plexus, which comprises the sinusoids of the pars distalis. This system brings the hypothalamic hormones into close proximity with the appropriate cell types in the pars distalis. For example, CTH-RF (corticotropin-releasing factor) is synthesized in the hypothalamus, released into the primary capillary plexus in the median eminence, and subsequently carried in the portal system to the secondary capillary plexus, where it interacts with corticotrophs in the pars distalis.

413. The answer is E. *(Junqueira, 8/e, pp 83, 146, 399, 404–409. Moore, Before We Are Born, 4/e, pp 159–160. Moore, Developing Human, 5/e, p 196. Ross, 3/e, pp 603–606, 612. Stevens, p 256–258.)* The thyroid gland functions under the action of the thyrotrophs in the anterior pituitary. The thyrotrophs, or thyroid follicular cells, import iodide and amino acids from the capillary lumen. The follicular cells synthesize thyroglobulin from the amino acids. When iodide enters the follicular cells, it undergoes oxidation. Thyroglobulin is iodinated while in the colloid and is the storage product of the thyroid follicular cells. The thyroid follicular cells pinocytose the iodinated thyroglobulin and the activity of lysosomes breaks down the colloid to form thyroxine (T_4), triiodothyronine (T_3), diiodotyrosine, and monoiodotyrosine. Most of the secretion of the human thyroid gland is composed of thyroxine, although triiodothyronine is more potent.

The thyroid gland also produces calcitonin, which is synthesized by the interfollicular cells. These cells are also known as parafollicular, or C, cells and are derived embryologically from the ultimobranchial bodies that form from the fourth pair of branchial pouches and perhaps a portion of the fifth pouch. Calcitonin decreases elevated serum calcium levels by transiently inhibiting osteoclastic activity.

414. The answer is B. *(Junqueira, 8/e, p 405. Ross, 3/e, pp 602–603, 609. Stevens, p 255.)* The photomicrograph illustrates the structure of the pineal gland, or epiphysis, which arises as an outgrowth of the diencephalon. The pineal contains two main cell types: pinealocytes and neuroglia (the latter appear to be modified astrocytes). The pinealocytes secrete melatonin in response to the light-dark cycle and influence the rhythmicity of other endocrine organs. In a sense the pineal therefore functions as a biologic clock. The pineal is innervated by postganglionic sympathetic fibers in a fashion similar to other glands in the head and neck region (e.g., salivary glands). The adrenal medulla is innervated by preganglionic sympathetic fibers. Corticotrophs in the pars distalis and possibly cells in the pars intermedia produce MSH. The pineal does not contain melanocytes or secrete MSH. There are age-related changes in the pineal in which the number of concretions and the degree of calcification of the "brain sand" increase. The pineal can be identified in x-rays by the calcification.

415. The answer is A. *(Junqueira, 8/e, pp 388–392. Ross, 3/e, pp 614, 616, 619. Stevens, pp 260–263.)* The female patient in this question suffers from adrenogenital or congenital virilizing hyperplasia. In this metabolic disease, there is a deficiency in the pathway that leads to cortisol synthesis. The inability to synthesize cortisol in turn leads to production of high levels of ACTH and ACTH releasing factor from the hypothalamus. The result is hypertrophy of the zona reticularis, which, like the zona fasciculata, is regulated by ACTH from the corticotrophs in the anterior pituitary. These cells produce the androgen dehydroepiandrosterone, which masculinizes the female genitals. Hypersecretion of cortisol cannot occur because of the metabolic defect in this pathway. The zona glomerulosa is regulated primarily by angiotensin II and only to a minor extent by ACTH.

416. The answer is A. *(Greenspan, 3/e, pp 254–260. Junqueira, 8/e, pp 146, 404–405. Ross, 3/e, pp 606–609, 613. Stevens, p 241.)* Parathyroid hormone (PTH), produced by the principal (chief) cells of the parathyroid gland, elevates blood calcium levels by stimulating ruffled border activity and increasing the number of osteoclasts. Receptors for PTH are found on osteoblasts, which secrete a mediator that subsequently stimulates osteoclastic activity. PTH also stimulates osteocytic osteolysis and Ca^{2+} transport by osteoblast-like bone-lining cells at the periosteum and endosteum. Calcium removed from the bone enters the bloodstream by way of the bone extracellular fluid, which results in an elevation of blood calcium. PTH increases excretion of PO_4^{2-} by the kidneys, which prevents elevated phosphate levels in the blood. It decreases Ca^{2+} excretion from the kidneys, but also increases absorption of Ca^{2+} from the small intestine, both of which increase blood calcium levels. PTH also stimulates the synthesis of dihydroxyvitamin D in the small intestine. Vitamin D regulates intestinal absorption of Ca^{2+} but also increases the differentiation of osteoclasts from monocyte precursors, which in both cases results in an increase in blood Ca^{2+} levels.

417. The answer is E. *(Junqueira, 8/e, pp 398, 401. Ross, 3/e, pp 603–606, 628–629. Stevens, pp 256–257.)* The thyroid gland is composed of follicles filled with colloidal material and surrounded by follicular cells with a cuboidal-to-columnar epithelium. Connective tissue separates the thyroid follicles. Cortisone is produced by the adrenal cortex; milk is produced by the lactating mammary gland; parathyroid hormone is synthesized by the parathyroid gland; and melanocyte-stimulating hormone (MSH) is synthesized by corticotrophs in the pars distalis. The thyroid, unlike the mammary gland, has no ducts.

418. The answer is D. *(Junqueira, 8/e, pp 378–383. Moore, Before We Are Born, 4/e, pp 292–293. Ross, 3/e, pp 596–597.)* The pituitary gland (hypophysis cerebri) is formed from ectoderm. An outgrowth of the oral ectoderm, Rathke's pouch, forms the structures that compose the adenohypophysis: pars distalis, pars intermedia, and pars tuberalis. The pars distalis includes the classic histologic cell types: chromophils (acidophils and basophils) and chromophobes (acidophils and basophils that are depleted of secretory product). Since the development of immunocytochemistry, the classification scheme for pars distalis cell types has been changed to include lactotrophs (prolactin), somatotrophs (growth hormone), corticotrophs (ACTH, β-lipotropin, α-MSH and β-endorphin), thyrotrophs (TSH), and gonadotrophs (FSH and LH). The pars intermedia is also formed from the oral ectoderm, is rudimentary in humans, and may produce a preproopiomelanocortic peptide. The pars tuberalis forms a collar around the pituitary stalk and is also derived from the oral ectoderm. The pars nervosa and the remainder of the pituitary stalk (infundibular stem and median eminence) are formed from a downgrowth of the diencephalon. The posterior pituitary (pars nervosa and stalk) retains this close relationship with the brain (i.e. hypothalamus) throughout life.

419. The answer is D. *(Junqueira, 8/e, pp 381–382, 399, 404–405. Ross, 3/e, pp 596–599, 604. Stevens, pp 251–254.)* The pituitary is known as the *master gland* and oversees the action of many other endocrine glands. The cells of the adenohypophysis produce hormones that regulate the functions of numerous organs and systems throughout the body. Release of the anterior pituitary hormones is controlled by negative feedback mediated by hypothalamic releasing hormones secreted by the hypothalamus. After hypophysectomy, loss of adrenocorticotropic hormone (ACTH) results in the lack of stimulation and reduced size of the glucocorticoid-producing cells of the adrenal zona fasciculata. Loss of thyrotropin (TSH) results in lack of stimulation and reduced activity of the thyroid follicular cells. Loss of follicle-stimulating hormone (FSH) results in failure of the maturation of the germ cells, so that spermatogenesis or growth of ovarian follicles ceases. In the case of benign pituitary adenomas, there is an excess production of anterior pituitary hormones with expected symptoms including an increase in the activities inhibited by hypophysectomy. The effects of the tumor vary depending on the pituitary cell types. The parathyroid chief cells synthesize parathyroid hormone (parathormone) in response to low serum calcium levels. Unlike many of the endocrine glands, the parathyroid gland appears to be independent of pituitary regulation.

420. The answer is C. *(Cotran, 5/e, pp 915–922. Greenspan, 3/e, pp 641–647. Isselbacher, 13/e, pp 1994–1997.)* Diabetes mellitus is a major health problem in the United States. Five percent of the population will develop diabetes in their lifetimes and 2.5 percent of Americans have diabetes. Complications of diabetes are microangiopathic (retinopathy and nephropathy that affect the small blood vessels); macroangiopathic (heart disease, strokes, and peripheral vascular disease, which may lead to gangrene); and neurologic (sensory, motor, and autonomic damage).

421–422. The answers are 421-A, 422-C. *(Junqueira, 8/e, pp 388–394. Ross, 3/e, pp 609–619. Stevens, pp 260–263.)* The adrenal cortex and the adrenal medulla may be considered two separate glands from a functional point of view. The adrenal cortex is derived embryologically from the coelomic epithelium and it is involved in steroid synthesis. Adrenocortical cells import cholesterol and acetate and produce androgens (zona reticularis), glucocorticoids (zona fasciculata), and mineralocorticoids (zona glomerulosa). The zona glomerulosa, found immediately beneath the capsule (E), is the site of the birth of new adrenocortical cells, which migrate through the zona fasciculata to reach the zona reticularis where cells die. Turnover in the adrenal cortex is slow compared to that which occurs in the gastrointestinal tract. Adrenocorticotropic hormone (ACTH), produced by basophils in the anterior pituitary, is the primary regulator of adrenocortical function. ACTH primarily affects the zona fasciculata, which produces the glucocorticoids, and the zona reticularis, which secretes androgens. ACTH stimulates the synthesis and secretion of mineralocorticoids (e.g., aldosterone) from the zona glomerulosa, but zona glomerulosa function is primarily regulated by angiotensin II. The adrenal medulla is derived from neural crest cells and it may be considered as modified postganglionic sympathetic fibers. Adrenal medullary cells synthesize and secrete norepinephrine, epinephrine, and enkephalins in response to stimulation of preganglionic sympathetic fibers that innervate the gland.

423–424. The answers are 423-C, 424-A. *(Burkitt, 3/e, pp 304–306, 308. Moore, Before We Are Born, 4/e, pp 290, 292–293. Moore, Developing Human, 5/e, pp 405–408. Ross, 3/e, pp 622–623. Stevens, pp 250–254.)* The pituitary is derived from the endoderm of the oral cavity (Rathke's pouch) and the floor of the diencephalon. The anterior (C) and intermediate (H) lobes and pars tuberalis (G) are derived from the oral cavity, while the remainder of the pituitary (pars nervosa [A] and the pituitary stalk [D]) is derived from a neuroepithelial origin. The cleft of Rathke's pouch (B) represents the lumen of the structure formed originally from the oral cavity. The pars distalis (C) contains acidophils and basophils regulated by stimulatory and inhibitory hormones produced by the hypothalamus. CRH is one of the adenohypophyseal regulatory factors synthesized in the hypothalamus, stored in the median eminence, and regulating corticotroph function. Factors such as CRH reach the anterior pituitary through the portal system. In the pars nervosa (A) the major cell type present is the pituicyte, a supportive glial cell. Axons that originate in the supraoptic and paraventricular nuclei descend into the pars nervosa. Oxytocin and vasopressin are stored in dilated endings in the pars nervosa called *Herring bodies.* These secretions are therefore synthesized in the hypothalamus and stored in the pars nervosa of the pituitary. Stucture E is the median eminence; F represents the cavity of the third ventricle.

425–427. The answers are 425-A, 426-C, 427-B. *(Junqueira, 8/e, pp 395– 398. Ross, 3/e, p 514. Stevens, p 264.)* The islets contain a number of cell types including the β-cells, which synthesize insulin; α-cells, which synthesize glucagon; δ-cells, which synthesize somatostatin; D1, or type IV, cells, which synthesize VIP; and F cells, which synthesize pancreatic polypeptide. Insulin reduces blood sugar levels by increasing cellular permeability to glucose; glucagon increases blood sugar levels and is secreted during hypoglycemia. Somatostatin regulates the release of insulin and glucagon. VIP is a vasodilator and inhibits gastric acid secretion. The function of pancreatic polypeptide is unknown.

Male Reproductive System

DIRECTIONS: Each question below contains five suggested responses. Select the **one best** response to each question.

428. Which is the correct order for spermatogenic cell differentiation within the seminiferous tubules?

(A) Round spermatid, elongating spermatid, testicular sperm, spermatocytes, spermatogonia
(B) Spermatocytes, round spermatid, elongating spermatid, testicular sperm, spermatogonia
(C) Spermatogonia, spermatocytes, round spermatid, elongating spermatid, testicular sperm
(D) Spermatocytes, spermatogonia, round spermatid, elongating spermatid, testicular sperm
(E) Spermatogonia, spermatocytes, elongating spermatid, round spermatid, testicular sperm

429. The primary regulator of Leydig cell secretion is

(A) follicle-stimulating hormone (FSH)
(B) luteinizing hormone (LH)
(C) FSH releasing factor
(D) inhibin
(E) androgen-binding protein

430. Sertoli cells are accurately described by which of the following statements?

(A) They divide during each wave of spermatogenic cell division
(B) They are found in a 1:1 relationship with spermatogonia, spermatocytes, and spermatids
(C) They fail to form junctional complexes with each other
(D) They are responsible for the synthesis of testosterone
(E) They provide protection, support, and nutrition for the development of the sperm

431. The formation of the acrosome

(A) occurs in the epididymis
(B) occurs after the release of the developing spermatids from the Sertoli cells
(C) involves the maturation of lysosomal enzymes
(D) involves mitotic activity
(E) involves meiotic divisions

Questions 432–434

Refer to the photomicrograph of the testis below.

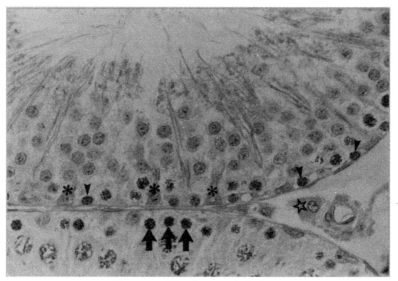

Courtesy of Dr. George C. Enders.

432. The cells labeled with arrows are

(A) primary spermatocytes
(B) secondary spermatocytes
(C) spermatids
(D) Sertoli cells
(E) spermatogonia

433. The cells labeled with arrowheads are

(A) primary spermatocytes
(B) secondary spermatocytes
(C) spermatids
(D) Sertoli cells
(E) spermatogonia

434. The cells marked with asterisks are

(A) primary spermatocytes
(B) secondary spermatocytes
(C) Leydig cells
(D) Sertoli cells
(E) spermatogonia

Questions 435–437

Refer to the photomicrograph from the reproductive system below.

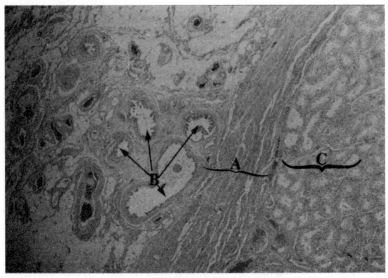

Courtesy of Dr. George C. Enders.

435. Identify the structure or structures labeled A.

(A) Seminiferous tubules
(B) Rete testis
(C) Corpus cavernosum of the penis
(D) Muscular wall of the oviduct
(E) Efferent ductules

436. Identify the structure or structures labeled B.

(A) Rete testis
(B) Efferent ductules
(C) Seminiferous tubules
(D) Vas deferens
(E) Oviduct

437. Identify the structure or structures labeled C.

(A) Ovary
(B) Epididymis
(C) Efferent ductules
(D) Straight tubules
(E) Seminiferous tubules

438. Identify the organ below.

Courtesy of Dr. George C. Enders.

(A) Male urethra
(B) Female urethra
(C) Ureter
(D) Vas deferens
(E) Epididymis

439. The portion of the prostate that is the most common site of prostatic cancer is the

(A) mucosal gland
(B) transition zone (submucosal gland)
(C) main gland
(D) central zone
(E) periurethral glands

440. Identify the organ below.

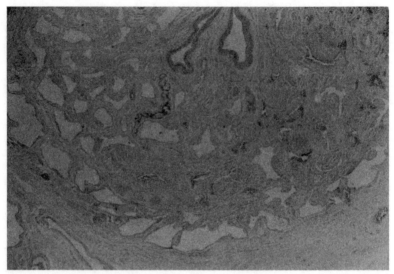

Courtesy of Dr. George C. Enders.

(A) Female urethra
(B) Male urethra
(C) Oviduct
(D) Ureter
(E) Seminal vesicle

441. Identify the organ below.

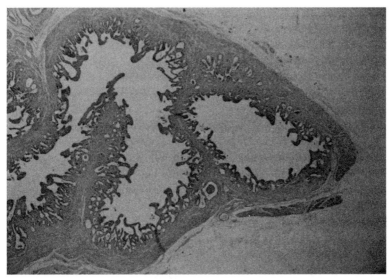

Courtesy of Dr. George C. Enders.

(A) Oviduct
(B) Seminal vesicle
(C) Prostate
(D) Efferent ductules
(E) Rete testis

442. The organ shown in this photomicrograph is responsible for production of

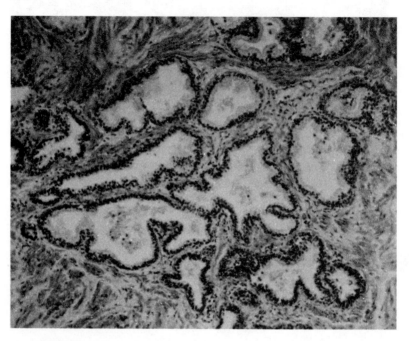

(A) fructose
(B) prostaglandins
(C) proteins that coagulate semen
(D) acid phosphatase
(E) ascorbic acid

DIRECTIONS: Each numbered question or incomplete statement below is NEGATIVELY phrased. Select the **one best** lettered response.

443. The organ shown in this photomicrograph is involved in all the following EXCEPT

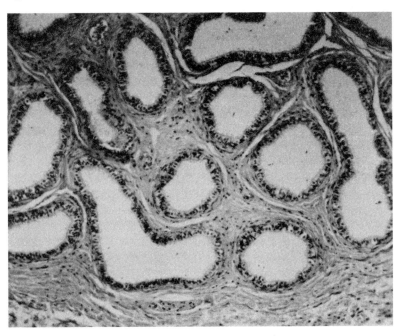

(A) spermatogenesis
(B) storage of sperm
(C) maturation of sperm
(D) phagocytosis of sperm
(E) absorption of testicular fluid

444. All the following are dependent on testosterone or other androgens EXCEPT

(A) the structural integrity of the prostatic epithelium
(B) the functional integrity of the prostatic glands
(C) development of the penis
(D) spermatogenesis
(E) fetal development of the testis from an indifferent gonad

445. Capacitation requires all the following EXCEPT

(A) a delay period before initiation
(B) loss of decapacitation factors
(C) a decrease in the fluidity of the plasma membrane
(D) release of the acrosomal enzymes
(E) fusion of the acrosomal membrane with the egg plasma membrane

446. All the following facts are related to the perception that sperm are immunologically foreign EXCEPT

(A) a blood-testis barrier exists
(B) genetic recombination of haploid sperm creates novel antigens
(C) cryptorchid testes are often incapable of producing fertile sperm
(D) vasectomy may lead to the production of antisperm antibodies
(E) vasovasostomy may not correct infertility

Male Reproductive System

Answers

428. The answer is C. *(Junqueira, 8/e, pp 407–413. Ross, 3/e, pp 636–639, 641–645, 649. Stevens, pp 306–309.)* The spermatogenic lineage begins with spermatogonia, which undergo spermatocytogenesis to form spermatocytes. Subsequently, the spermatocytes undergo meiosis and produce the haploid number associated with gametes. The differentiation of spermatocytes to sperm is called *spermiogenesis*. Spermatogonia are of three types: type A dark cells (Ad), type A pale cells (Ap), and type B cells. Type Ad cells serve as precursors with each division resulting in the formation of Ad and Ad or Ad and Ap cells. The Ap cells give rise to B cells, which are capable of differentiating into primary spermatocytes by passage through meiosis. After the first meiotic division the spermatogenic cells are known as *secondary spermatocytes,* which exist for a short period of time before they enter the second meiotic division to form spermatids. The spermatids begin as round structures and elongate with the formation of a flagellum. At the completion of spermiogenesis, mature sperm are released into the lumen of the seminiferous tubules.

429. The answer is B. *(Junqueira, 8/e, pp 413–417. Ross, 3/e, pp 654–655. Stevens, p 320.)* Luteinizing hormone (LH), also known as *interstitial cell–stimulating hormone (ICSH),* regulates the production of testosterone by Leydig cells. Testosterone is modified by binding to androgen-binding protein (ABP), which is synthesized by the Sertoli cells. The testosterone is necessary for the maintenance of spermatogenesis as well as the male ducts and accessory glands. The Sertoli cells also produce inhibin, which feeds back on the anterior pituitary and hypothalamus to regulate FSH release. ABP is regulated by FSH, testosterone, and inhibin.

430. The answer is E. *(Junqueira, 8/e, pp 413–415. Ross, 3/e, pp 638, 647–648, 650, 652–654. Stevens, pp 311–312.)* Sertoli cells are supportive cells that serve a nutritional function for the developing sperm, phagocytose spermatid fragments, and secrete several important factors as well as fluid, which facilitates the movement of sperm toward the epididymis and male duct system. Sertoli cells produce androgen-binding protein, inhibin, ceruloplasmin (a copper-binding protein), transferrin (an iron-binding protein), and müllerian-inhibiting substance, but not testosterone, which is produced by the Leydig cells in the connective tissue spaces between the seminiferous tubules. The base of the Sertoli cell is located outside the blood-testis barrier, while the apical portion lies luminal to the blood-testis barrier. Gap junctions between Sertoli cells provide communication between adjacent cells. During spermatogenesis, derivatives of spermatocytes cross from the basal to the adluminal compartment across the zonula occludens between adjacent Sertoli cells. Each Sertoli cell is associated with multiple spermatogenic cells.

431. The answer is C. *(Junqueira, 8/e, pp 409–413. Stevens, pp 308–309.)* The formation of the acrosome is one stage in spermiogenesis, the process by which mature sperm are formed from the spermatids. The process occurs after the division of secondary spermatocytes. It involves no mitotic or meiotic activity and can be separated into three or four stages depending upon the system used for classification. The phases include (1) Golgi, (2) cap, (3) acrosome, and (4) maturation phases, although many textbooks omit the cap stage. In the Golgi phase, the acrosomal vesicle forms a cap over the anterior half of the nucleus. The acrosome phase involves the maturation of the lysosomal enzymes (including acrosin, a serine protease, hyaluronidase, and neuraminidase), which will be responsible for the penetration ability of the sperm. In the maturation phase, cytoplasm is pinched off and subsequently phagocytosed by Sertoli cells. The developing cells are in contact with the Sertoli cells through the third and fourth stages of spermiogenesis. At the end of spermiogenesis, spermatids are released by the Sertoli cells in a process called *spermiation*.

432–434. The answers are 432-A, 433-E, 434-D. *(Burkitt, 3/e, pp 325–326, 330. Ross, 3/e, pp 639, 666–667.)* The testis is composed of seminiferous tubules composed of a number of cell types. Most of these cells are involved in spermatogenesis and spermiogenesis and are part of the spermatogenic series. The Sertoli cells function in a nutritive and supportive role somewhat analogous to the glial cells of the CNS. The cells labeled with the arrowheads are spermatogonia, the derivatives of the embryonic primordial germ cells. These cells compose the basal layer and undergo mitosis to form primary spermatocytes, which have distinctive clumped or coarse chromatin (marked by arrows). Secondary spermatocytes are formed during the first meiotic division and exist for only a short period of time because there is no lag period before entry into the second meiotic division. Sertoli cells are the supportive cells labeled with asterisks in the figure. There are extensive tight junctions between them. The cell marked with a star is a Leydig cell. It is located between seminiferous tubules and is responsible for the production of testosterone.

435–437. The answers are 435-B, 436-B, 437-E. *(Burkitt, 3/e, pp 330–333, 344. Ross, 3/e, pp 655–658, 668–669.)* The photomicrograph is taken from an area that shows the ductuli efferentes (efferent ductules) (B), the seminiferous tubules (C), and the mediastinum testis containing the rete testis (A). Sperm leave the seminiferous tubules through short tubuli recti into the straight tubules of the rete testis. The efferent ductules have a distinctive wavy epithelium in which adjoining cells are tall and ciliated and short and nonciliated.

438. The answer is D. *(Burkitt, 3/e, p 331. Ross, 3/e, pp 660, 672–673.)* The organ shown in the figure is the vas deferens (ductus deferens). The vas deferens conducts sperm from the epididymis to the urethra. The thick muscular wall is unique in the presence of an inner longitudinal, a middle circular, and an outer longitudinal layer of smooth muscle.

439. The answer is C. *(Cotran, 5/e, pp 1023–1028. Junqueira, 8/e, pp 419–420. Ross, 3/e, pp 656–658. Stevens, pp 315–317.)* The prostate consists of three parts: (1) a small mucosal (inner periurethral) gland, (2) a transition zone that consists of a submucosal (outer periurethral) gland, and (3) a peripheral portion known as the *main,* or *external, gland.* Most adenocarcinoma of the prostate occurs in the main gland and is often undiagnosed until the later symptoms of back pain or blockage of the urethra are detected. Benign prostatic hypertrophy, also known as *benign nodular hyperplasia,* occurs in the mucosal and submucosal glands, which are rarely the sites of inflammation or carcinoma. The main gland is sensitive to androgens while the periurethral glands are sensitive to androgens and estrogens. Acid phosphatase levels may be valuable in the diagnosis of prostatic carcinoma and its metastasis.

440. The answer is B. *(Burkitt, 3/e, pp 333–334. Junqueira, 8/e, pp 420–421. Ross, 3/e, p 665. Stevens, p 318.)* The male urethra possesses a primarily pseudostratified columnar type of epithelium. The exceptions are the prostatic urethra, which has a transitional epithelium, and pockets of stratified squamous epithelium that may occur throughout the urethra. The thick-walled arteries of the penile and cavernous sinuses of the penile erectile tissue are also a distinguishing feature of this organ. The helicine arteries supply the sinuses. Action of the parasympathetic nervous system mediates the dilatation of these vessels during erection.

441. The answer is B. *(Burkitt, 3/e, pp 332, 344. Ross, 3/e, pp 676–677, 720–721.)* The seminal vesicle produces about 50 percent of the seminal fluid on a volume basis. Its secretion consists of fructose, prostaglandins, and vitamin C. The wall is composed of muscle and the mucosa of anastomosing villi-like structures.

442. The answer is D. *(Burkitt, 3/e, pp 323, 332–333. Erlandsen, pp 150, 160. Junqueira, 8/e, pp 419–420. Ross, 3/e, pp 659–664. Stevens, pp 314–317.)* The organ shown in the light microscopic photograph is the prostate. The prostate is composed of 15 to 30 tubuloalveolar glands surrounded by fibromuscular tissue. Concretions are often found in the lumina. The prostate secretes a thin, opalescent fluid that contributes primarily to the first part of the ejaculate. Prostatic secretions include acid phosphatase, spermine (a polyamine), fibrolysin, amylase, and zinc. Spermine oxidation results in the musky odor of semen and fibrolysin is responsible for the liquefaction of semen after ejaculation. Acid phosphatase along with prostatic-specific antigen are important for the diagnosis of metastases. Fructose, prostaglandins, proteins responsible for semen coagulation, and ascorbic acid are part of the seminal fluid, which is produced by the seminal vesicle and makes up most of the ejaculate.

443. The answer is A. *(Burkitt, 3/e, p 331. Erlandsen, pp 149, 158. Junqueira, 8/e, pp 417– 419. Stevens, p 313.)* The figure is a light microscopic photograph of the epididymis. The epididymis functions in the storage, maturation, and phagocytosis of sperm and residual bodies. In addition, the epididymis is involved in the absorption of testicular fluid and the secretion of glycoproteins. These glycoproteins may be involved in the inhibition of capacitation. The epithelium of the epididymis is pseudostratified with stereocilia (modified microvilli for absorption), and the wall contains extensive smooth muscle. Sperm are often found in the lumina.

444. The answer is E. *(Junqueira, 8/e, p 416. Moore, Before We Are Born, 4/e, pp 213–215. Moore, Developing Human, 5/e, pp 282–283. Stevens, pp 319–320.)* During fetal development, the production of androgens by the developing testis results in masculinization of the indifferent gonadal ducts and the indifferent genitalia. In the absence of androgens, female genitalia and female ducts (vagina, oviducts, and uterus) develop. The development of the testis from an indifferent gonad is dependent on the presence of the testis-determining factor, a gene on the short arm of the Y chromosome. In the mature male, testosterone is required for the initiation and maintenance of spermatogenesis as well as the structural and functional integrity of the accessory glands and ducts of the male reproductive system. Testosterone is bound to androgen-binding protein (ABP), which is synthesized by the Sertoli cells under the influence of follicle-stimulating hormone (FSH). ABP is important for both the storage and delivery of androgens in the male ducts and accessory glands.

445. The answer is C. *(Junqueira, 8/e, pp 410, 431. Moore, Before We Are Born, 4/e, pp 26–27. Moore, Developing Human, 5/e, p 28. Stevens, p 310.)* Capacitation is a process that prepares the sperm for fertilization. Sperm must reside in the female reproductive tract or under appropriate in vitro conditions for about 1 h in order for capacitation to occur. During this delay phase there is a loss of decapacitation factors, which have been added to the sperm by Sertoli and epididymal cells. Cholesterol is removed from the plasma membrane during this period, which results in the increased fluidity of the membrane that is required for the fusion of the acrosomal membrane with the egg plasma membrane. This leads to the release of the acrosomal enzymes, which are required for the breakdown of the corona radiata and the zona pellucida of the oocyte to facilitate sperm penetration.

446. The answer is C. *(Isselbacher, 13/e, pp 2006–2010. Ross, 3/e, p 652.)* Sperm are immunologically foreign because of a number of factors. Spermatogenesis begins at puberty long after the development of self-recognition in the immune system. The blood-testis barrier protects developing sperm from exposure to systemic factors. The basal compartment containing the spermatogonia and preleptotene spermatocytes is exposed to plasma; however, the adluminal compartment, which contains primary and secondary spermatocytes, spermatids, and testicular sperm, prevents these antigens from entering the blood. In the case of vasectomy, sperm that has leaked from the severed vas deferens is viewed as foreign by immune surveillance and antibodies develop. The phagocytosis of sperm by macrophages plays a role in the development of antisperm antibodies that occurs following the ligation or removal of a segment of the vas deferens. Attempted reunion of the ligated segments is called *vasovasostomy* and may return sperm to the ejaculate; however, the presence of antisperm antibodies may prevent normal fertilization. The inability of cryptorchid testes to produce fertile sperm is related to the higher temperature in the abdomen than in the normal scrotal location.

Female Reproductive System

DIRECTIONS: Each question below contains five suggested responses. Select the **one best** response to each question.

447. Elevated estrogen levels

(A) decrease LH levels
(B) increase secretion of plasminogen activator and collagenase
(C) increase FSH levels
(D) decrease the ciliation of the epithelial cells of the oviduct
(E) decrease basal body temperature at the time of ovulation

448. Which of the following are characteristic of the secretory phase of the menstrual cycle?

(A) It precedes ovulation
(B) It depends on progesterone secretion by the corpus luteum
(C) It coincides with the development of ovarian follicles
(D) It coincides with a rapid drop in estrogen levels
(E) It produces ischemia and necrosis of the stratum functionale

449. Secretion of plasminogen activator and collagenase by granulosa cells is required for

(A) dissolution of the zona pellucida to facilitate sperm penetration
(B) pH regulation within the antral cavity
(C) breakdown of the basement membrane between the thecal and granulosa layers, facilitating ovulation
(D) diffusion of androgens between the thecal and granulosa cells
(E) facilitation of follicular atresia through breakdown of the basement membrane between the theca interna and externa

450. The low pH in the vagina is maintained by

(A) a proton pump similar to that of parietal cells and osteoclasts
(B) acid secretion derived from intracellular carbonic acid
(C) secretion of lactic acid by the stratified squamous epithelium
(D) bacterial metabolism of glycogen to form lactic acid
(E) synthesis and accumulation of acid hydrolases in the epithelium

451. Primary oocytes have developed by the time of birth. From puberty to menopause, these germ cells remain suspended in meiotic prophase. The oocyte of a mature follicle is induced to undergo the first meiotic division just prior to ovulation as a result of which of the following hormonal stimuli?

(A) The cessation of progesterone secretion
(B) The gradual elevation of follicle-stimulating hormone (FSH) titers
(C) The low estrogen titers associated with the maturing follicle
(D) The slow elevation of progesterone produced by luteal cells
(E) The surge of luteinizing hormone (LH) initiated by high estrogen titers

452. The secondary oocyte enters the second meiotic division and proceeds as far as metaphase. The stimulation for continuation of the second meiotic division to produce the haploid ovum is

(A) elevation of progesterone titers
(B) the environment of the oviduct and uterus
(C) expulsion from the mature follicle
(D) fertilization by a spermatozoon
(E) the presence of human chorionic gonadotropin (hCG)

453. A 33-year-old patient with an average menstrual cycle of 28 days comes in for a routine Pap smear. It has been 35 days since the start of her last menstrual period, and a vaginal smear reveals clumps of basophilic cells. You suspect

(A) she will begin menstruating in a few days
(B) she will ovulate within a few days
(C) her serum progesterone levels are very low
(D) there are detectable levels of hCG in her serum and urine
(E) she is undergoing menopause

Questions 454–456

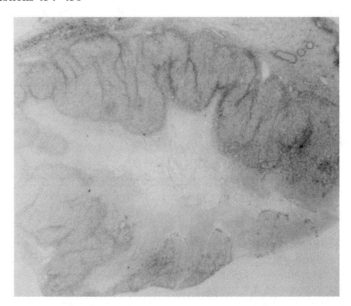

454. The photomicrograph above illustrates

(A) a corpus albicans
(B) a corpus luteum
(C) placenta
(D) adrenal gland
(E) hypophysis

455. The structure in the photomicrograph is formed and maintained primarily in response to _____ and _____ , respectively.

(A) LH; LH or human chorionic gonadotropin (hCG)
(B) FSH; prolactin
(C) FSH; LH or FSH
(D) Prolactin; LH
(E) Estrogen; progesterone

456. If the necessary stimulating hormone was absent 12 to 14 days after ovulation in a human female, the result would be

(A) the absence of the structure
(B) the absence of muscularization
(C) maintenance of the uterine epithelium for implantation beyond 14 days after ovulation
(D) pregnancy
(E) the formation of a corpus albicans from the structure

457. The accompanying diagram shows a cross section of a developing human endometrium and myometrium. Hormonal ratios control the development of which of the labeled vessels?

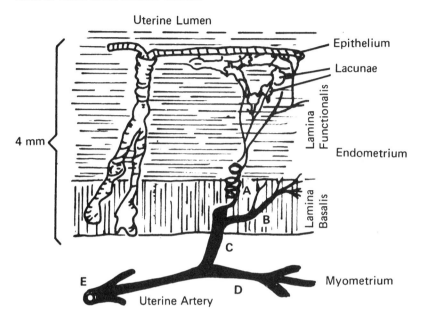

(A) A
(B) B
(C) C
(D) D
(E) E

458. The chorionic villi shown in the accompanying photomicrograph are derived from

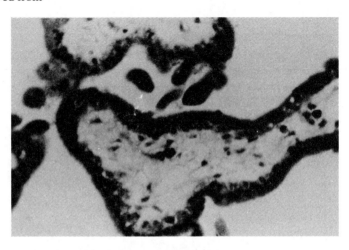

(A) a combination of fetal and maternal tissues
(B) endometrial glands
(C) endometrial stroma
(D) fetal tissues
(E) maternal blood vessels

---

459. Synthesis of milk by the mammary gland requires

(A) oxytocin
(B) production of prolactin by the corpus luteum
(C) the influence of vasopressin
(D) holocrine secretion of milk proteins
(E) hormonal stimulation of alveolar proliferation

460. Carcinoma of the breast

(A) is usually alveolar in origin
(B) has a high mortality because of the absence of cytotoxic T cells in the breast
(C) rarely metastasizes
(D) is rare compared to other cancers in females
(E) can be detected early by regular mammography

461. Which of the following statements is true of the uterus during the menstrual cycle?

(A) Progesterone secretion initiates proliferation in the endometrium

(B) Estrogen secretion stimulates secretory changes in the endometrium

(C) Cessation of estrogen and progesterone secretion results in the degeneration of the endometrium

(D) Incorporation of ^{3}H-thymidine in the uterus of an experimental animal would occur primarily during the secretory phase

(E) Studies with an antibody to the estrogen receptor would demonstrate a peak in immunocytochemically positive endometrial cells after involution of the corpus luteum

462. Naturally occurring, nonpathologic cervical eversions ("erosions") are usually naturally corrected by reepithelialization. These eversions are most prevalent in which one of the following reproductive classifications of women?

(A) Prepubertal female

(B) Postpubertal, premenopausal, nulliparous female

(C) Premenopausal, multiparous female

(D) Menopausal, nulliparous female

(E) Late postmenopausal female

DIRECTIONS: The group of questions below consists of lettered headings followed by a set of numbered items. For each numbered item select the **one** lettered heading with which it is **most** closely associated. Each lettered heading may be used **once, more than once, or not at all.**

Questions 463–465

Match each description with the appropriate label on the diagram of an ovarian follicle.

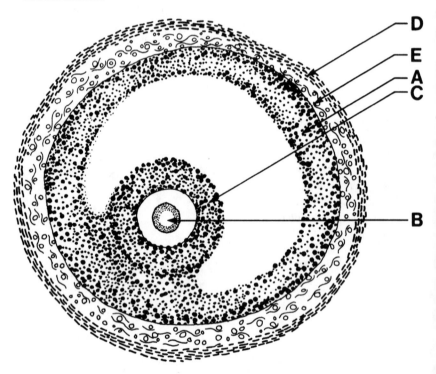

463. Cells in this layer are involved in the synthesis of androstenedione

464. These cells remain in close contact with the zona pellucida

465. These cells convert androgens to estradiol

Female Reproductive System

Answers

447. The answer is B. *(Junqueira, 8/e, pp 427–431.)* Estrogen levels increase during the maturation of ovarian follicles, which results in a decrease in FSH and causes an LH surge. Elevated estrogen levels also result in increased secretion of lytic enzymes, prostaglandins, plasminogen activator, and collagenase to facilitate the rupture of the ovarian wall and the release of the ovum and the attached corona radiata. Following ovulation, during the luteal phase of the cycle, the theca and granulosa cells are transformed into the corpus luteum under the influence of LH. Increased ciliation and height of the oviductal lining cells are also under the influence of estrogen. Increases in the number of cilia serve to facilitate movement of the ovum. Ovulation occurs near the middle of the menstrual cycle. There is an increase in basal body temperature at the time of ovulation that appears to be indirectly regulated by elevated estrogen levels. Estrogen also influences the appearance of FSH receptors on granulosa cell membranes and enhances synthesis and storage of glycogen in the vaginal epithelium.

448. The answer is B. *(Burkitt, 3/e, pp 339–341. Junqueira, 8/e, pp 432–436. Ross, 3/e, pp 694–700.)* The secretory phase of the menstrual cycle follows the proliferative (follicular) phase and is followed by the menstrual phase. During the follicular phase (approximately days 4 to 16) estrogen produced by the ovaries drives cell proliferation in the base of endometrial glands and the uterine stroma. The proliferative phase culminates with ovulation. The secretory phase (approximately days 16 to 25) is characterized by high progesterone levels from the corpus luteum, a tortuous appearance of the uterine glands, and apocrine secretion by the gland cells. During this phase, maximum endometrial thickness occurs. The menstrual phase (approximately days 26 to 30) is characterized by decreased glandular secretion and eventual glandular degeneration because of decreased production of both progesterone and estrogen by the theca lutein cells. Contraction of coiled arteries and arterioles leads to ischemia and necrosis of the stratum functionale.

449. The answer is C. *(Burkitt, 3/e, pp 337, 340. Junqueira, 8/e, pp 427–430. Ross, 3/e, pp 685–686, 690–691.)* Substances that facilitate rupture of the ovarian follicle during ovulation include prostaglandins, plasminogen activator, plasmin, collagenase, and luteinizing hormone (LH). The increase in LH in midcycle induces production of collagenase and plasminogen activator. These proteases facilitate ovulation by initiating connective tissue remodeling, including the breakdown of the basement membrane between thecal and granulosa layers. Connective tissue remodeling is involved in the process of follicular atresia. This process occurs throughout life and involves the death of follicular cells as well as oocytes, but there is no basement membrane between the theca interna and externa. In fact, there is an absence of a clear delineation between the theca interna and externa.

450. The answer is D. *(Burkitt, 3/e, p 353. Junqueira, 8/e, p 439. Ross, 3/e, p 708. Stevens, p 325.)* The vagina has a stratified squamous epithelium that contains large accumulations of glycogen. Glycogen is released into the vaginal lumen and is subsequently metabolized to lactic acid by commensal lactobacilli. The low pH inhibits growth of a variety of microorganisms. Treatment for vaginal infections usually includes acidified carriers to reestablish a more acidic pH like that usually seen in mid-menstrual cycle.

451. The answer is E. *(Alberts, 3/e, pp 1022–1024. Junqueira, 8/e, pp 425, 429–430. Ross, 3/e, pp 680–683, 686–688. Stevens, pp 24–25.)* Follicle-stimulating hormone (FSH) and luteinizing hormone (LH) produced in the adenohypophysis result in growth and maturation of the ovarian follicle. Under FSH stimulation, the theca cells proliferate, hypertrophy, and begin to produce estrogen. A midcycle surge of LH appears to trigger the resumption of meiosis and to cause the FSH-primed follicle to rupture and discharge the ovum. Under the influence of LH, the ruptured follicle is transformed to a corpus luteum, which produces progesterone.

452. The answer is D. *(Alberts, 3/e, pp 1022–1024. Junqueira, 8/e, pp 425, 429–430. Ross, 3/e, pp 686–688. Stevens, pp 24–25.)* The secondary oocyte enters the second meiotic division just prior to ovulation and arrests at metaphase. Fertilization by a spermatozoon provides the stimulation for the division of chromatin to the haploid number. By the time the fertilized ovum reaches the uterus, the progesterone produced by the corpus luteum has initiated the secretory phase in the endometrium. Once implantation occurs and the chorion develops, human chorionic gonadotropin (hCG) is synthesized and the corpus luteum is maintained.

453. The answer is D. *(Burkitt, 3/e, p 352. Junqueira, 8/e, pp 440–442. Yen, 3/e, pp 837–839.)* The patient described in this question is probably pregnant. The delay in menstruation coupled with the presence of basophilic cells in a vaginal smear is a clue. Ovulation is the midpoint of the cycle and should be more than a few days away. She is relatively young for the onset of menopause and there are no other symptoms. The vaginal epithelium varies little with the normal menstrual cycle. Exfoliative cytology can be used to diagnose cancer and to determine if the epithelium is under stimulation of estrogen and progesterone. The presence of basophilic cells in the smear with the Pap staining method would indicate the presence of both estrogen and progesterone. The data suggest the maintenance of the corpus luteum (i.e., pregnancy).

454–456. The answers are 454-B, 455-A, 456-E. *(Burkitt, 3/e, pp 341–342. Erlandsen, pp 161, 166. Junqueira, 8/e, pp 429–430. Ross, 3/e, pp 688–690.)* The structure in the photomicrograph is a corpus luteum. The corpus luteum forms from the granulosa and theca layers of the follicle following ovulation. The luteal phase is the second half of the menstrual period and follows the follicular phase during which follicles mature. The corpus luteum synthesizes progesterone in response to high LH levels. In each reproductive cycle, the production of LH stimulates development and maintenance of the corpus luteum, which is well formed by 12 to 14 days following ovulation. In the case of fertilization and subsequent implantation, the corpus luteum of pregnancy is maintained by human chorionic gonadotropin (hCG) produced by the embryo. In the absence of LH or hCG, the corpus luteum regresses to form a corpus albicans, which consists primarily of fibrous connective tissue. Without LH or hCG, the uterine epithelium, which has undergone glandular proliferation in preparation for implantation, undergoes collapse and degeneration as part of menstruation.

457. The answer is A. *(Burkitt, 3/e, pp 346–349. Erlandsen, pp 162, 169. Junqueira, 8/e, pp 434–435. Ross, 3/e, pp 693–699. Stevens, pp 340–341.)* The spiral arteries of the endometrium (labeled A in the diagram accompanying the question) are dependent on specific estrogen/progesterone ratios for their development. They pass through the basalis layer of the endometrium into the functional zone, and their distal ends are subject to degeneration with each menses. The straight arteries (B) are not subject to these hormonal changes. In the proliferative phase the endometrium is only 1 to 3 mm thick, and the glands are straight, with the spiral arteries only slightly coiled. This diagram of the early secretory phase shows an edematous endometrium that is 4 mm thick, with glands that are large, beginning to sacculate in the deeper mucosa, and coiled for their entire length. In the late secretory phase, the endometrium becomes 6 to 7 mm thick.

458. The answer is D. *(Junqueira, 8/e, pp 436–440. Moore, Before We Are Born, 4/e, pp 29–33. Moore, Developing Human, 5/e, pp 33–37. Ross, 3/e, pp*

728–731. Stevens, pp 344–346.) A fertilized ovum reaches the uterus about 4 days after fertilization. At that time, it has developed into a multicellular, hollow sphere referred to as a *blastocyst*. The blastocyst soon adheres to the secretory endometrium and differentiates into an inner cell mass that will develop into the embryo and a layer of primitive trophoblast. The expanding trophoblast penetrates the surface endometrium and erodes into maternal blood vessels. Eventually, it develops two layers, an inner cytotrophoblast and an outer syncytiotrophoblast. Solid cords of trophoblast form the chorionic villi, which then are invaded by fetal blood vessels.

459. The answer is E. *(Junqueira, 8/e, pp 443–446. Ross, 3/e, pp 709–713. Stevens, pp 366–367. Yen, 3/e, pp 373–376.)* The mammary gland enlarges during pregnancy in response to several hormones including prolactin synthesized by the anterior pituitary, estrogen and progesterone synthesized by the corpus luteum, and placental lactogen. The alveoli at the end of the duct system respond to these hormones by cell proliferation, which increases the size of the mammary glands. Growth continues throughout pregnancy; however, secretion is most notable late in pregnancy. Milk is synthesized in the alveoli and is stored in their lumina before passage through the lactiferous ducts to the nipples. Secretion of milk lipids occurs by an apocrine mechanism whereby some apical cytoplasm is included with the secretory product. In comparison, milk proteins such as the caseins are secreted by exocytosis. Oxytocin is required for the release of milk from the mammary gland through the action of the myoepithelial cells that surround the alveoli and proximal (closer to the alveolus) portions of the duct system. Oxytocin is not required for milk synthesis. Neurohumoral reflexes are involved in the suckling–milk ejection response.

460. The answer is E. *(Cotran, 5/e, pp 1099–1102. Junqueira, 8/e, p 446. Stevens, p 368.)* Carcinoma of the breast occurs in about 1 of 10 females in the United States. By definition a carcinoma is ductal in origin. Carcinoma of the breast metastasizes to the brain, lungs, and bones. The easy access of tumor cells to the extensive axillary blood supply and lymphatic drainage facilitates the spread of the cancer into the blood and lymph supplies. Self-examination and mammography are urged in an attempt to increase early diagnosis, which has reduced mortality of this disease.

461. The answer is C. *(Burkitt, 3/e, pp 346–349. Junqueira, 8/e, pp 432–436. Stevens, p 341.)* During the uterine cycle menstruation is initiated by the necrosis of the stratum functionale through the action of the spiral arterioles. After 4 to 5 days, proliferation begins in the endometrium in response to estrogen secretion from the granulosa cells. If ^{3}H-thymidine were injected into animals during this period and combined with immunocytochemical staining for antiestrogen receptor, the result would be colocalization of autoradiographic grains and immunocytochemical product over the uterine epithelial and stromal cells as well as vascular endothelial cells. The peak of proliferative activity and estrogen sensitivity would occur after menses and during the proliferative phase. Maintenance of cell proliferation requires continued secretion of estrogen. As progesterone secretion increases, the secretory phase of uterine maturation occurs. The involution of the corpus luteum initiates the degeneration of the endometrial glands, which precedes menses.

462. The answer is C. *(Cotran, 5/e, pp 1045–1046. Ross, 3/e, 697–700, 726–727. Stevens, p 328.)* To a minor extent, the uterine cervical stroma changes during each reproductive cycle; however, during pregnancy (especially parturition) there is a thinning of the uterine stroma. This results in eversions (mistakenly called "erosions"), which are sites of exposed uterine columnar epithelium in the acidic, vaginal milieu. These sites often become reepithelialized as stratified epithelium (squamous metaplasia) and are believed to be the location of cancerous transformation in the cervix. As part of the process of reepithelialization, the openings of cervical mucous glands are obliterated, which results in the formation of cysts called *nabothian follicles.*

463–465. The answers are 463-E, 464-C, 465-A. *(Burkitt, 3/e, pp 338–340. Erlandsen, pp 261, 263. Junqueira, 8/e, pp 423–426. Ross, 3/e, pp 680–685, 714–717.)* The diagram illustrates a mature (graafian) follicle. Development of ovarian follicles begins with a primordial follicle that consists of flattened follicular cells surrounding a primary oocyte. During the follicular phase these cells undergo mitosis to form multiple granulosa layers (primary follicle) in response to elevated levels of FSH and LH from the anterior pituitary. A glycoproteinaceous coat surrounds the oocyte and is called the *zona pellucida.* The connective tissue around the follicle differentiates into two layers: theca externa (D) and theca interna (E). The theca externa is closest to the ovarian stroma and consists of a highly vascular connective tissue. The theca interna synthesizes androgens (e.g., androstenedione) in response to LH. Androgens are converted to estradiol by the action of an aromatase enzyme synthesized by the granulosa cells under the influence of FSH. Increased levels of estrogen from the ovary feed back to decrease FSH secretion from gonadotropes in the anterior pituitary. Liquor folliculi is produced by the granulosa cells and is secreted between the cells. When cavities are first formed by the development of follicular fluid between the cells, the follicle is called *secondary.* When the antrum is completely formed, the follicle is called a *ma-*

ture (graafian) follicle and the antrum is completely filled with liquor folliculi. The granulosa cells form two structures. The corona radiata (C) represents those granulosa cells that remain attached to the zona pellucida. The cumulus oophorus (not labeled) represents those granulosa cells that surround the oocyte (B) and connect it to the wall. Structure A is the membrana granulosa.

Urinary System

DIRECTIONS: Each question below contains five suggested responses. Select the **one best** response to each question.

466. In the accompanying transmission electron micrograph of the renal corpuscle, the cell marked with an asterisk is

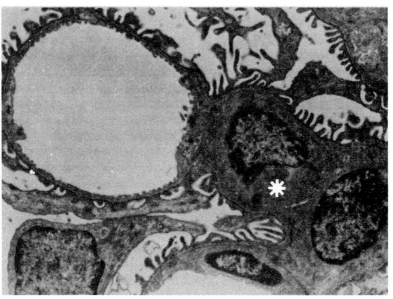

Courtesy of Dr. Vincent H. Gattone

(A) a mesangial cell
(B) an endothelial cell
(C) a podocyte
(D) a visceral epithelial cell
(E) a parietal cell

467. In the accompanying transmission electron micrograph from the renal corpuscle, the structures labeled with arrows are

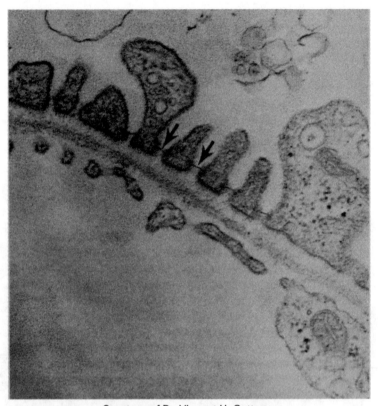

Courtesy of Dr. Vincent H. Gattone

(A) podocyte foot processes
(B) endothelial cell fenestrations
(C) pedicels
(D) filtration slits
(E) the lamina rara of the basement membrane

468. The accompanying transmission electron micrograph from the kidney illustrates a cell from the

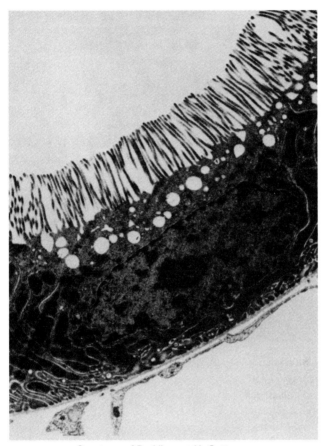

Courtesy of Dr. Vincent H. Gattone.

(A) proximal convoluted tubule
(B) distal convoluted tubule
(C) collecting duct
(D) vasa recta
(E) descending thin limb

469. The arrows in the accompanying scanning electron micrograph of the renal glomerulus indicate

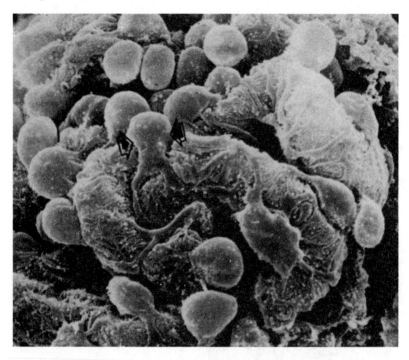

(A) a red blood cell
(B) a podocyte
(C) an endothelial cell
(D) a parietal cell
(E) the macula densa

470. Changes in the glomerular filtration barrier in diabetes mellitus include

(A) thinning of the basement membrane
(B) hyperalbuminemia
(C) dehydration
(D) proteinuria
(E) enlargement of the filtration slits between foot processes

DIRECTIONS: Each numbered question or incomplete statement below is NEGATIVELY phrased. Select the **one best** lettered response.

471. All the following are involved in the glomerular filtration barrier EXCEPT

(A) physical barrier function
(B) charge barrier
(C) type IV collagen
(D) filtration slits between foot processes of adjacent podocytes
(E) a positive charge in the membrane due to the presence of heparan sulfate

472. All the following are found exclusively in the renal cortex EXCEPT

(A) proximal convoluted tubules
(B) distal convoluted tubules
(C) collecting ducts
(D) afferent arterioles
(E) glomeruli

DIRECTIONS: The group of questions below consists of lettered headings followed by a set of numbered items. For each numbered item select the **one** lettered heading with which it is **most** closely associated. Each lettered heading may be used **once, more than once, or not at all.**

Questions 473–476

Match each description with the appropriate structure.

(A) Proximal tubule
(B) Thin limb of the loop of Henle
(C) Distal tubule and thick ascending limb
(D) Collecting duct
(E) Afferent arterioles
(F) Vasa recta
(G) Peritubular capillary network
(H) Mesangial cells
(I) Podocytes

473. Impermeability to water despite the presence of ADH

474. Site of countercurrent multiplier

475. Site of action of aldosterone

476. Capillaries involved in maintenance of hyperosmotic medulla

Urinary System
Answers

466. The answer is A. *(Junqueira, 8/e, pp 359, 368. Ross, 3/e, pp 567–568. Stevens, pp 281–282.)* The transmission electron micrograph illustrates a renal corpuscle. The cell labeled is a mesangial cell, which is morphologically similar to pericytes found in association with other systemic blood vessels. The mesangial cells surround glomerular capillaries as illustrated in this electron micrograph. They are proposed to have a number of potential functions including phagocytosis, regulation of glomerular blood flow, and maintenance of the basement membrane of the glomerulus. A podocyte with its processes in close association with the glomerular capillary is observed below the mesangial cell. The podocytes form the visceral layer of the Bowman's capsule. The outer layer of the Bowman's capsule is formed by parietal cells, one of which is located in the lower left corner of the micrograph. An endothelial cell within a glomerular capillary is also shown below the mesangial cell.

467. The answer is D. *(Junqueira, 8/e, pp 359, 363–367. Stevens, pp 278–280.)* The transmission electron micrograph illustrates the filtration barrier of the renal corpuscle. The structures labeled are the filtration slits, which are located between adjacent pedicels (foot processes of the podocytes). The remainder of the filtration barrier is formed by the glomerular basement membrane, which contains type IV collagen and heparan sulfate. There are three distinct layers within the glomerular basement membrane: (1) an electron-dense lamina densa in the center surrounded by (2) the lamina rara externa on the glomerular side and by (3) the lamina rara interna on the capillary endothelial side.

468. The answer is A. *(Junqueira, 8/e, pp 359–360, 362–366, 369–371. Ross, 3/e, pp 571–575. Stevens, pp 284–289.)* The transmission electron micrograph illustrates a proximal convoluted tubule cell. This cell type is responsible for reabsorption from the glomerular filtrate. The elaborate microvilli at the apical surface increase the surface area available for absorption. The apical portion of the cells also possesses an extensive endocytic vacuolar arrangement for the reabsorption of proteins. Numerous mitochondria are located basally to provide energy for the reabsorption of water that follows sodium transported actively by a basolateral Na+,K+-ATPase. In the distal tubule there are very few microvilli. The distal tubule pumps ions against ionic gradients and is responsive to aldosterone. The collecting duct epithelium contains many mitochondria, fewer microvilli, and no basal infoldings as occur in the cells of the proximal convoluted tubule. The vasa recta are lined by endothelial cells and the thin loops show some resemblance to the capillaries.

469. The answer is B. *(Junqueira, 8/e, pp 362, 364–365. Ross, 3/e, pp 568–569. Stevens, p 280.)* The arrows on the scanning electron micrograph illustrate a podocyte. The podocytes surround the glomerular capillaries. The spaces between the foot processes (pedicels) form the filtration slits, an important part of the filtration barrier of the kidney. The macula densa is a portion of the distal tubule that is specialized for determination of distal tubular osmolarity.

470. The answer is D. *(Junqueira, 8/e, p 364. Ross, 3/e, pp 566–569. Stevens, pp 279–280.)* In diabetes mellitus, the glomerular basement membrane is thickened. The separation of laminae rarae and densa is obliterated, which results in a loss of selectivity of the filtration barrier. This causes the loss of protein from the blood to the urine (proteinuria). The liver adjusts to the proteinuria by producing more proteins (e.g., albumin). After continued proteinuria, the liver is unable to produce sufficient protein, which results in hypoalbuminemia and edema of the tissues. The foot processes are affected in many diseases that lead to nephrotic syndrome. Loss of anionic charge and fusion of the foot processes result in the obliteration of the filtration slits.

471. The answer is E. *(Junqueira, 8/e, pp 359–364, 366–367. Ross, 3/e, pp 562–569. Stevens, pp 278–280.)* The glomerular filtration barrier is a physical and charge barrier that exhibits selectivity based on molecular size and charge. The barrier is formed by three components: (1) glomerular capillary endothelial cells, (2) glomerular basement membrane, and (3) podocyte layer. The presence of collagen type IV in the lamina densa of the basement membrane presents a physical barrier to the passage of large proteins from the blood to the urinary space. Glycosaminoglycans, particularly heparan sulfate, produce a polyanionic charge that binds cationic molecules. Filtration slits are found between adjacent podocyte foot processes and provide a gap of approximately 50 μm. The foot processes are coated with a glycoprotein called *podocalyxin*, which is rich in sialic acid and provides mutual repulsion to maintain the structure of the filtration slits. It also possesses a large polyanionic charge for repulsion of large anionic proteins.

472. The answer is C. *(Junqueira, 8/e, pp 361, 369, 371, 375. Ross, 3/e, pp 558–563. Stevens, p 283.)* The collecting ducts are found in both the cortex and medulla of the kidney. Cortical collecting ducts are found in the medullary rays, while medullary collecting ducts are found in the medulla and lead into the papillary duct.

473–476. The answers are 473-C, 474-B, 475-C, 476-F. *(Junqueira, 8/e, pp 359–375. Ross, 3/e, pp 558–579. Stevens, pp 273–277, 281–289.)* The function of the kidney is the formation of urine from the filtration of the plasma. Under hormonal regulation, the kidney will modify the concentration of the urine through the reabsorption of ions, water, carbohydrates, and proteins of small molecular weight. The kidney is divided into two parts: an outer cortex and an inner medulla.

Blood supply is through the renal artery, which enters the hilum of the kidney and distributes blood to the interlobar arteries, which travel between the medullary pyramids. At the corticomedullary junction, the interlobar arteries distribute blood to the arcuate arteries, which lead to the interlobular arteries and ultimately to the afferent arterioles. The afferent arterioles are associated with the juxtaglomerular cells, which are modified arterial smooth muscle cells that produce important substances involved in the regulation of blood pressure. Blood from the afferent arteriole enters the glomerular capillaries, which are responsible for the filtration of the plasma. The glomerular capillaries sit on mesangial cells and extracellular mesangial matrix. The mesangial cells provide support but also contain myosin and possess angiotensin II receptors on their surface. The glomerulus is surrounded by the Bowman's capsule, which is a double-layered epithelial capsule. The Bowman's capsule consists of two layers: an outer parietal layer and an inner visceral layer that contains the podocytes, the epithelial cells that make a major contribution to the glomerular filtration barrier.

Following the glomerulus, the blood in most areas of the kidney enters the efferent arterioles and then the peritubular capillary network. In the deep cortex, efferent arterioles empty into vasa recta, which travel into the medulla. In this case, the vasa recta play an important role in electrolyte and fluid exchange in the renal medulla. Blood in the peritubular capillary network is in close proximity to the cortical tubules and is able to pick up substances reabsorbed by the tubular epithelial cells. Lack of oxygenated blood in the peritubular capillary network leads to loss of cortical cell function, and if severe it may lead to acute tubular necrosis.

The nephron is the functional unit of the kidney with two parts associated with the two capillary networks. The glomerulus is associated with the glomerular capillaries while the cortical and medullary tubules are associated with the peritubular capillary networks and the vasa recta, respectively.

The tubular system begins with the proximal tubule (in the cortex), which is an extension of the Bowman's capsule. In order, the tubular system includes proximal convoluted tubule, proximal straight tubule (thick descending limb of the loop of Henle), thin limb of the loop of Henle (descending and ascending in the medulla), distal straight tubule (thick ascending limb of Henle back into the cortex), and distal convoluted tubule. The distal convoluted tubules empty into the collecting tubules, which drain into the minor calyces.

The proximal tubule possesses a well-developed brush border with numerous, long, regular microvilli on its apical surface. The combination of the proximal convoluted tubule and the proximal straight tubule is the longest nephron segment and extends from the superficial cortex to the deep medulla. It is the primary site for the reduction of the tubular fluid volume, and the presence of the extensive microvillus border reflects this function.

The thin loop of Henle is responsible for the production of the countercurrent multiplier, which allows the kidneys to produce a hyperosmotic medulla (i.e., 1600 mOsm versus isosmolar conditions of 290 mOsm). Through the regulation of collecting duct permeability, this allows the production of hyperosmotic urine. The multiplier increases NaCl in the medullary interstitial fluid. This occurs through the movement of Na^+ and Cl^- out of the ascending limb (which is impermeable to water) and into the interstitial fluid. Subsequently, the descending limb, which is permeable to water, takes up the Na^+ and Cl^- from the interstitium. The vasa recta are involved in countercurrent exchange; that is, they do not produce the hyperosmotic medulla but modify their osmolarity to that of the medulla during the course of the blood flow.

The distal tubule has the highest concentration of Na^+,K^+-ATPase in any segment of the nephron. It is impermeable to water despite the presence of antidiuretic hormone (ADH). The distal tubule is the site of action for aldosterone. This hormone is synthesized by the zona glomerulosa of the adrenal cortex and stimulates reabsorption of Na^+ and water from the glomerular filtrate. The distal tubules empty into the collecting ducts, which are permeable to water under the regulation of ADH. The collecting ducts in turn drain into the minor calyces.

Eye

DIRECTIONS: Each question below contains five suggested responses. Select the **one best** response to each question.

477. In the surgical procedure known as *radial keratotomy,* slits are made in the cornea to flatten it slightly. This will result in

(A) decreased refraction of light by the cornea
(B) a decreased amount of light entering through the cornea
(C) conversion of the cornea from a "stationary" to an "adjustable" form of refraction
(D) maintenance of the lens in a more flattened state
(E) focusing of light on the retina at a point other than the fovea

478. Retinal detachment most commonly results from

(A) local swelling in specific retinal layers
(B) leakage of blood from the inner retinal capillaries
(C) fluid accumulation between the retina and the retinal pigment epithelium (RPE)
(D) impaired pumping of water toward the photoreceptors by the retinal pigment epithelium
(E) increased phagocytosis of outer segments by the retinal pigment epithelial cells

DIRECTIONS: Each numbered question or incomplete statement below is NEGATIVELY phrased. Select the **one best** lettered response.

479. Visual transduction involves all the following EXCEPT

(A) activation of phosphodiesterase
(B) a decrease in levels of cGMP
(C) conversion of 11-*cis* retinal to all-*trans* retinal
(D) opening of an Na⁺ channel
(E) a decrease in the release of neurotransmitter by the photoreceptor

480. The retinal pigmented epithelium (RPE) performs all the following functions EXCEPT

(A) regulation of ion and metabolite transport between the blood and the retina as part of the blood-retinal barrier
(B) containment of the photosensitive cells
(C) storage of vitamin A in the form of a fatty ester for later use in the photoreceptor cells
(D) synthesis of melanin
(E) phagocytosis of worn-out components of the photoreceptor cells

481. Diabetic retinopathy may be characterized by all the following EXCEPT

(A) thickening of the basal lamina of small retinal vessels
(B) microaneurysms
(C) hemorrhage of retinal vessels
(D) retinal ischemia and proliferation of new vessels
(E) loss of phagocytotic ability of the pigmented epithelium

DIRECTIONS: Each group of questions below consists of lettered headings followed by a set of numbered items. For each numbered item select the **one** lettered heading with which it is **most** closely associated. Each lettered heading may be used **once, more than once, or not at all.**

Questions 482–485

For each of the following descriptions, match the corresponding letter of the structure in the diagram.

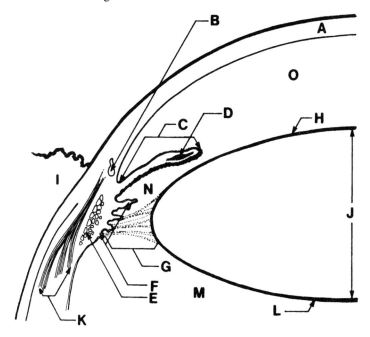

482. Similarity to microfibrils of elastic fibers and involvement in accommodation

483. Responsibility for production of aqueous humor

484. Site of outflow of aqueous humor

485. Blockage of this structure results in glaucoma

Questions 486–489

Match each description with the appropriate structure.

(A) Meibomian glands
(B) Lacrimal glands
(C) Lens epithelium
(D) Choroid
(E) Dilator muscle of the iris
(F) Müller's cells
(G) Amacrine cells
(H) Outer segment of the rod
(I) Inner segment of the rod
(J) Cones
(K) Inner nuclear layer of the retina
(L) Outer plexiform layer of the retina

486. Secretion helps decrease evaporation of tear film on the corneal surface

487. Source of blood supply to the photoreceptors

488. Data from photoreceptor cells are integrated

489. Light is detected by this segment of the photoreceptor

Eye

Answers

477. The answer is A. *(Newell, 7/e, pp 221–222. Vaughan, 13/e, pp 147, 429.)* Radial, or refractive, keratotomy alters corneal (not lens) anatomy to create a new shape that is more flattened in the center and higher at the periphery of the cornea. This occurs because intraocular pressure will cause a reshaping of the cornea due to the induced weakness produced by the incisions. The purpose of the procedure is to reduce myopia in order to eliminate the need for corrective lenses. The reduction in curvature of the central portion of the cornea results in decreased refractive power of the cornea. The slits may be made with a laser and the degree of correction required may be estimated by computer simulations.

478. The answer is C. *(Newell, 7/e, pp 310–313. Vaughan, 13/e, pp 185–186, 200–201, 430.)* Retinal detachment is the result of the accumulation of fluid between the retina and the RPE. In one type of detachment, rhegmatogenous retinal detachment, fluid accumulates after a break occurs in the retina. Detachments without breaks in the retina are called *nonrhegmatogenous,* or *serous, detachments.* Vitreous degeneration usually is a prerequisite for retinal detachment that results in the breaking of the retina. The breakdown of the vitreous produces traction on the retina, which may already possess an inherent area of weakness. The site of retinal detachment is the area between the inner and outer layers of the embryonic optic cup and represents a relatively weak area of adherence between the retinal and RPE layers.

479. The answer is D. *(Junqueira, 8/e, pp 461–463. Ross, 3/e, pp 751–752. Stevens, p 199.)* Rhodopsin is the visual pigment of rod cells and is composed of retinal, a vitamin A derivative, bound to opsins. Photons reaching rhodopsin isomerize retinal to the all-*trans* form from 11-*cis* retinal. The result is bleaching, which represents the dissociation of retinal from the opsins. The bleaching process results in a fall in cGMP within the cytosol. Transducin is a G protein that couples bleaching to cGMP through the action of a phosphodiesterase enzyme that cleaves cGMP to GMP. The closing of the Na+ channel results in a reduction in permeability to sodium ions and hyperpolarization of the cell membrane. The signal spreads to the inner segment and through gap junctions to nearby photoreceptor cells. In the presence of cGMP, the Na+ channel remains open; in its absence, the channel closes and the cell hyperpolarizes. Therefore, the rods and cones differ from other receptors in that hyperpolarization of the cell membranes occurs rather than the depolarization that occurs in other neural systems. Closing the channel slows down the release of the visual transmitter.

480.   The answer is B. *(Junqueira, 8/e, pp 456–459. Ross, 3/e, pp 741–742, 749–752. Stevens, p 198.)* The retinal pigment epithelium (RPE) is a single layer of cells derived from the outer layer of the optic cup. It is continuous from the ora serrata retinae to the optic nerve. Microvilli are prominent on the apical surfaces of the RPE and play an important role in the maintenance of the blood-retinal barrier. In addition, the RPE synthesizes melanin, phagocytoses old components of the photoreceptor cells, and stores vitamin A, which is used by the photoreceptor cells.

481.   The answer is E. *(Junqueira, 8/e, pp 461–463. Newell, 7/e, pp 492–500. Stevens, p 201. Vaughan, 13/e, pp 203–206.)* The ophthalmoscope is extremely valuable in the analysis of vascular changes in diseases such as hypertension and diabetes mellitus. In diabetes mellitus one of the major complications is diabetic retinopathy. In diabetic retinopathy the pathologic changes usually begin with thickening of the basement membrane of the small retinal vessels. The abnormal vessels develop microaneurysms, which leak and hemorrhage with resultant ischemia of the retinal tissue. New vessels proliferate in response to the ischemia and the production of angiogenic factors. Loss of phagocytosis by the RPE occurs in retinal dystrophy but is not a characteristic of diabetic retinopathy.

482–485.   The answers are 482-G, 483-F, 484-B, 485-B. *(Junqueira, 8/e, pp 449–456. Ross, 3/e, pp 740–749, 760–761, 764–767. Stevens, pp 193–197.)* The figure shows the region of the iridocorneal angle and other associated structures in the eye. This region is extremely important in the production and outflow of the aqueous humor and in the distribution of zonule fibers to the lens.

The eye has three layers and three compartments. The outermost layer of the eye consists of the fibrous sclera in the posterior four-fifths and the cornea in the anterior one-fifth. The middle layer is the vascular uvea. There are three components of the uvea. The choroid comprises the posterior four-fifths, the iris is the most anterior portion, and the ciliary body is the intermediate part of the uvea. The innermost layer is composed of the derivatives of the optic cup: the pigmented epithelium and the retina itself. The three compartments of the eye are the anterior and posterior chambers and the vitreous body.

The iris (C) contains both sphincter (D) and dilator muscles, which work in opposition to one another and are innervated by parasympathetic and sympathetic fibers, respectively. They form the anterior portion of the iris, while the posterior part is formed by the retinal layers. The center of the iris is a hole—the pupil.

The ciliary body contains the ciliary muscles (E and K) and tissue derived from the retinal layers. The ciliary muscle stretches the choroid and relaxes the lens, which is essential for the process of lens accommodation. The ciliary processes extend from the ciliary body and produce aqueous humor. They are also the origin of the zonule fibers (G), which are similar in structure to elastic fibers and are important for anchorage of the lens. The zonule fibers are involved in accommodation. When the ciliary muscles contract, causing forward displacement of the ciliary body, the tension on the zonule fibers is reduced, which leads to an increase in lens thickness and maintenance of focus.

The aqueous humor produced by the ciliary processes (F) is transported into the posterior chamber (N) and flows into the anterior chamber (O). Outflow from the anterior chamber occurs through the trabecular meshwork at the iridocorneal angle and flows through the sinus venosus sclerae (canal of Schlemm, B) to the scleral veins and the systemic vasculature. Blockage of the canal of Schlemm, the trabecular meshwork, or the scleral veins results in glaucoma.

The cornea (A) forms the transparent, avascular anterior portion of the eye. It is similar in structure to the sclera, but has a higher content of glycosaminoglycans. The cornea possesses a mechanism to prevent turgescence in which fluid is continually pumped from the corneal tissue by the endothelial cells. The outer anterior surface of the cornea is covered by an epithelium. Beneath the epithelium is Bowman's membrane, the corneal stroma, Descemet's membrane, and the endothelium (at the posterior surface of the cornea), which lines the anterior boundary of the anterior chamber.

The lens (J) is formed embryologically from a thickening of the surface ectoderm called the *lens placode,* which eventually forms a lens vesicle. Lens fiber production continues throughout life with no turnover. The lens is surrounded by a capsule and an underlying epithelium (H and L).

The three compartments of the eye include the posterior and anterior chambers, which are filled with aqueous humor, and the vitreous body (M), which is filled with a gel consisting of hydrated hyaluronic acid and other glycosaminoglycans.

The conjunctiva is the mucosa, or lining, of the eyelid and is labeled I in the figure.

486–489. The answers are 486-A, 487-D, 488-K, 489-H. (*Junqueira, 8/e, pp 449–466. Ross, 3/e, pp 749–756, 762–763. Stevens, pp 195–205.*) The retina consists of nine layers:

1. The photoreceptor layer consisting of the rods and cones is the outer layer of the retina with the photoreceptor cells protruding across the external (outer) limiting membrane.

2. The outer limiting membrane is formed by the junctional complexes between Müller's cells and the membranes of photoreceptor cells.

3. The outer nuclear layer contains the nuclei and rod and cone perikarya.

4. The outer plexiform layer contains rod and cone synapses as well as the cell processes of bipolar, horizontal, and photoreceptor cells.

5. The inner nuclear (bipolar) layer is composed of the nuclei and perikarya of the bipolar and amacrine cells as well as the nuclei of Müller's cells. This layer is responsible for the integration of data from adjacent photoreceptors.

6. The inner plexiform layer consists of amacrine cells dispersed between the processes of bipolar and ganglion cells. The interplexiform cells are also found in this layer. These cells extend between the processes of bipolar and ganglion cells and synapse between the photoreceptor cells in the outer plexiform layer. The inner plexiform layer is responsible for modulation of signals from the ganglion to the photoreceptor cells.

7. The ganglion cell layer contains the ganglion cells separated by the cytoplasm of Müller's cells. The Müller's cells are astrocyte-like glia.

8. The nerve fiber layer consists of axons of the ganglion cells, which will form the optic nerve.

9. The internal limiting membrane is located between the vitreous and the retina.

The photoreceptors are of two types: rods and cones. The nuclei of the rods and cones are found in the outer nuclear layer and extend across the outer limiting membrane in one direction and toward the outer plexiform layer in the other direction. The outer segment is the photon-sensitive portion of the rod and cone and contains disks. Rhodopsin is composed of opsin and *cis*-retinal. It is responsible for transduction of light (photons) into hyperpolarization of the cell membrane. Rhodopsin is present in the disks of the outer segment of the rod. In the rods, there is constant turnover of the disks, which can be monitored by autoradiographic studies of the formation and shedding of disks in various animals. The RPE functions in the phagocytosis of rod disks. The inner segment contains numerous mitochondria, glycogen, and protein synthetic apparatus. Rods are responsible for night vision while the cones are responsible for color vision, which is best resolved at the fovea. The fovea, which is the center of the macula, is composed exclusively of cones and is the site of optimal resolution.

The eyelids contain a number of different glands all named by eponyms: meibomian glands, glands of Zeis, and glands of Moll. The meibomian glands produce a secretion rich in lipid that inhibits evaporation of tear film; therefore, these sebaceous glands serve to protect and lubricate the cornea.

The lacrimal glands produce the tears, which are a serous fluid produced by the acini of the main and accessory glands. The secretion passes through the ducts and into the conjunctiva. Drainage of tears occurs by way of ducts, the lacrimal sac, and eventually the nasolacrimal duct.

Proteins called *crystallins* are found within the lens and are formed by differentiation of the lens epithelial cells.

The choroid is a highly vascular layer that consists of three parts: stroma, choriocapillaris, and Bruch's membrane. Blood supply to the retina is derived from the choriocapillaris of the choroid.

The dilator muscles of the pupil are derived from neuroepithelial cells and are pigmented. There is no separation of muscle and epithelial layers; the cells are described as *pigmented myoepithelial cells.* The dilator of the pupil is innervated by sympathetic nervous system fibers.

Ear

DIRECTIONS: Each question below contains five suggested responses. Select the **one best** response to each question.

490. The malleus, incus, and stapes

(A) are formed from the first two branchial pouches
(B) are composed of hyaline cartilage in the adult
(C) undergo dampened motion through the action of the stapedius and tensor tympani
(D) constitute the inner ear
(E) are the mechanoreceptors of the ear

491. The direction in which vestibular hair cell stereocilia are deflected is important because

(A) it differentiates between type 1 and type 2 hair cells
(B) it determines whether cells are depolarized or hyperpolarized
(C) it determines whether linear or angular acceleration is detected
(D) it determines the direction of blood flow in the stria vascularis
(E) it is determined by the frequency of the sound

492. The middle ear is accurately characterized by which of the following statements?

(A) It contains perilymph
(B) It transmits sound to the round window through the auditory ossicles
(C) It contains the auditory ossicles, which serve to overcome the impedance of the fluids of the inner ear
(D) It is separated from the outer ear by Reissner's membrane
(E) It is connected to the inner ear by the helicotrema

DIRECTIONS: Each numbered question or incomplete statement below is NEGATIVELY phrased. Select the **one best** lettered response.

493. All the following are directly involved in sound transmission EXCEPT

(A) release of neurotransmitter onto the afferent endings of cranial nerve VIII
(B) shearing motion of tectorial membrane against hair cell stereocilia
(C) movement of the basilar membrane
(D) equalization of the pressure in the middle ear and the nasopharynx by the eustachian tube
(E) vibration of the tympanic membrane

494. Endolymph is located in all the following structures EXCEPT the

(A) utricle
(B) saccule
(C) semicircular canals
(D) scala media
(E) scala tympani

DIRECTIONS: Each group of questions below consists of lettered headings followed by a set of numbered items. For each numbered item select the **one** lettered heading with which it is **most** closely associated. Each lettered heading may be used **once, more than once, or not at all.**

Questions 495–497

For each structure listed below, select the labeled site on the diagram of the organ of Corti.

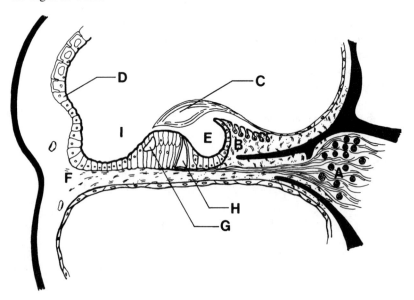

495. Tectorial membrane

496. Stria vascularis

497. Spiral ligament

Questions 498–500

Match each morphological or functional characteristic with the correct cell type or ear structure.

   (A) Maculae of the utricle and saccule
   (B) Semicircular canals
   (C) Interdental cells
   (D) Organ of Corti
   (E) Pillar cells

498. Detection of angular acceleration

499. Production of the tectorial membrane

500. Presence of otoconia (statoconia)

Ear

Answers

490. The answer is C. *(Junqueira, 8/e, pp 466–467. Stevens, pp 188–191.)* The malleus, incus, and stapes are the auditory ossicles and are derived embryologically from the first two branchial arches. They are composed of compact bone in the adult and their motion is dampened by the action of the stapedius and tensor tympani muscles as a protective action when someone is exposed to a loud noise. The mechanoreceptors of the inner ear are the hair cells of the inner ear.

491. The answer is B. *(Kandel, 3/e, pp 503–508. Ross, 3/e, pp 774–788.)* The vestibular hair cells are the sensory transduction system of the inner ear and are responsible for the conversion of mechanical energy into an electrical signal for cranial nerve VIII. These cells are called *hair cells* because their surface contains stereocilia. These are modified microvilli that contain a large number of actin filaments and extend from the surface of the cell. The stereocilia are different lengths and are arranged in order by size with a large kinocilium at one end. The arrangement of the stereocilia is very important as bending in one direction (i.e., toward the kinocilium) depolarizes the cell and increases the rate of nervous discharge, while bending them in the other direction (i.e., away from the kinocilium) results in hyperpolarization and decreased neural discharge. The type of hair cells is based on the pattern of efferent and afferent innervation.

492. The answer is C. *(Junqueira, 8/e, pp 466–470. Ross, 3/e, pp 768–774. Stevens, p 188.)* The middle ear contains the auditory ossicles, which transmit sound to the oval window and therefore serve in the transduction of sound waves to the perilymph. The helicotrema represents the opening that allows communication of the tympanic and vestibular cavities. The vestibular membrane, also known as *Reissner's membrane,* maintains the gradient between the endolymph of the scala media and the perilymph of the vestibular cavity. The epithelium possesses extensive occluding junctions, which serve to maintain the concentration gradient that is essential for sensory transduction.

493. The answer is D. *(Junqueira, 8/e, pp 470–472. Ross, 3/e, pp 768–769, 781–783, 788–789. Stevens, pp 190–191.)* Sound waves are directed toward the tympanic membrane by the pinna and the external auditory canal of the external ear. The vibration of the tympanic membrane is transmitted to the oval window by way of the ossicles of the middle ear. Induction of waves in the perilymph results in the movement of the basilar and vestibular membranes toward the scala tympani and causes the round window to bulge outwards. The movement of the hair cells is facilitated because the tectorial membrane is rigid and the pillar cells form a pivot. Shearing of the hair cell stereocilia against the tectorial membrane results in depolarization, and the release of neurotransmitter onto afferent endings of the auditory cranial nerve results in the initiation of an action potential. The stabilization of the pressure between the middle ear and the nasopharynx is not directly related to the mechanism of sound transmission.

494. The answer is E. *(Junqueira, 8/e, pp 468–470. Ross, 3/e, pp 771, 777, 782–783, 792. Stevens, pp 190–191.)* Endolymph is similar to intracellular fluid (high K+, low Na+). It is found in the utricle, saccule, semicircular canals, and scala media (cochlear duct), which are parts of the membranous labyrinth. Endolymph is synthesized by the highly vascular stria vascularis in the lateral wall of the scala media. The endolymphatic sac and duct are responsible for absorption of endolymph and the endocytosis of molecules from the endolymph. The vestibule contains perilymph, not endolymph.

495–497. The answers are 495-C, 496-D, 497-F. *(Burkitt, 3/e, pp 388, 393, 395. Junqueira, 8/e, pp 467–472. Stevens, pp 190–191.)* The organ of Corti is found within the cochlear duct and contains the hair cells that are responsible for transduction of the sound to a nerve impulse. It rests on the basilar membrane, which separates it from the epithelial lining of the tympanic cavity. The inner tunnel (H) of the organ of Corti separates the outer from the inner hair cells. The outer hair cells possess microvilli, which are attached to the tectorial membrane (C). In contrast, the inner hair cells are unattached. Supportive cells include the phalangeal and pillar cells, which are not labeled on the figure. The stria vascularis (D) is found in the lateral wall of the cochlear duct (scala media) (I). The spiral lamina is a bony structure that protrudes from the modiolus. The spiral limbus (B) is a connective tissue structure superior to the unattached edge of the spiral lamina. Along the outer wall of the canal of the organ of Corti is a thickened projection of periosteum known as the *spiral ligament* (F). The spiral ganglion is labeled A on the figure and contains bipolar cells. Peripheral processes of spiral ganglion cells reach the organ of Corti while central processes terminate in nuclei located in the medulla. The internal spiral tunnel is labeled E in the figure.

498–500. The answers are 498-B, 499-C, 500-A. *(Junqueira, 8/e, pp 467–472. Moore, Before We Are Born, 4/e, pp 308–310. Moore, Developing*

Human, 5/e, pp 433, 435. Ross, 3/e, pp 774–784. Stevens, pp 191–192.) The utricle represents the dorsal portion of the otocyst-derived inner ear; the saccule represents the ventral portion. Both the utricle and saccule contain maculae that detect linear acceleration. The maculae of the utricle and saccule are perpendicular to one another. These maculae contain type I and type II hair cells, which differ in their innervation. The hair cells have stereocilia and a kinocilium embedded in a membrane that contains otoconia (statoconia) composed of calcium carbonate. The semicircular canals, which extend from the utricle, contain the cristae ampullares and detect angular acceleration. The stereocilia and kinocilia are embedded in the cupola, which does not contain the otoconia found in the maculae. The endolymph turns right when the head turns left and vice versa. Movement stimulates the stereocilia and induces depolarization. The interdental cells produce the tectorial membrane, which is essential for the development of the shearing force in the process of sound transduction in the organ of Corti. It detects sound vibration and is responsive to variation in the frequency of sound waves.

Bibliography

Alberts B, Bray D, Lewis J, et al: *Molecular Biology of the Cell,* 3/e. New York, Garland, 1994.

Avery JK (ed): *Oral Development and Histology,* 2/e. New York, Thieme, 1994.

Budka H: Human immunodeficiency virus (HIV)-induced disease of the central nervous system: Pathology and implications for pathogenesis. *Acta Neuropathol* 77:225–236, 1989.

Burkitt HG, Young B, Heath JW: *Wheater's Functional Histology,* 3/e. New York, Churchill Livingstone, 1993.

Coe FL, Favus MJ (eds): *Disorders of Bone and Mineral Metabolism.* New York, Raven, 1992.

Cotran RS, Kumar V, Robbins SL: *Robbins' Pathologic Basis of Disease,* 5/e. Philadelphia, Saunders, 1994.

Donowitz M, Sharp GWG (eds): *Mechanism of Intestinal Electrolyte Transport and Regulation by Calcium.* New York, Liss, 1984.

Erlandsen SL, Magney JE: *Color Atlas of Histology.* St. Louis, Mosby, 1992.

Favus MJ (ed): *Primer on the Metabolic Bone Diseases and Disorders of Mineral Metabolism,* 2/e. Kelseyville, CA, American Society of Bone and Mineral Research, 1993.

Fawcett DW: *The Cell,* 2/e. Philadelphia, Saunders, 1981.

Fawcett DW: *A Textbook of Histology,* 11/e. Philadelphia, Saunders, 1986.

Gilbert SF: *Developmental Biology,* 4/e. Sunderland, MA, Sinauer, 1994.

Goodman SR (ed): *Medical Cell Biology.* Philadelphia, Lippincott, 1994.

Greenspan FS: *Basic and Clinical Endocrinology,* 3/e. East Norwalk, CT, Appleton & Lange, 1991.

Guyton AC: *Textbook of Medical Physiology,* 8/e. Philadelphia, Saunders, 1991.

Isselbacher KJ, Martin JB, Braunwald E, et al (eds): *Harrison's Principles of Internal Medicine,* 13/e. New York, McGraw-Hill, 1994.

Ito S, Lacy ER: Morphology of rat gastric mucosal damage, defense, and restitution in the presence of ethanol. *Gastroenterology* 88:250–260, 1985.

Jilka RL, Hangoc G, Girasole G, et al: Increased osteoclast development after estrogen loss: Mediation by interleukin-6. *Science* 257:88–91, 1992.

Johnson LR (ed): *Physiology of the Gastrointestinal Tract,* 3/e. New York, Raven, 1994.

Junqueira LC, Carneiro J, Kelley RO: *Basic Histology,* 8/e. East Norwalk, CT, Appleton & Lange, 1995.

Kandel ER, Schwartz JH, Jessell TM: *Principles of Neural Science,* 3/e. New York, Elsevier, 1991.

Larsen WJ: *Human Embryology.* New York, Churchill Livingstone, 1993.

Lebenthal E: *Human Gastrointestinal Development.* New York, Raven, 1989.

Male D: *Immunology: An illustrated outline,* 2/e. St. Louis, Mosby, 1991.

Mayne R, Burgeson RE (eds): *Structure and Function of Collagen Types.* New York, Academic, 1987.

Moore KL: *Clinically Oriented Anatomy,* 3/e. Baltimore, Williams & Wilkins, 1992.

Moore KL, Persaud TVN: *Before We Are Born,* 4/e. Philadelphia, Saunders, 1993.

Moore KL, Persaud TVN: *The Developing Human,* 5/e. Philadelphia, Saunders, 1993.

Newell FW: *Ophthalmology: Principles and Concepts,* 7/e. St. Louis, Mosby–Year Book, 1992.

Paul WE (ed): *Fundamental Immunology,* 3/e. New York, Raven, 1993.

Roitt I, Brostoff J, Male D: *Immunology,* 3/e. St. Louis, Mosby, 1993.

Ross MH, Romrell LJ, Kaye GI: *Histology: A Text Atlas,* 3/e. Baltimore, Williams & Wilkins, 1995.

Rubin W, Ross LL, Sleisenger MH, Jeffries MB: The normal human gastric epithelia. *Lab Invest* 19:598–626, 1968.

Simionescu M, Simionescu N, Palade G: Biochemically differentiated microdomains of the cell surface of the capillary endothelium. *Ann N Y Acad Sci* 401:9–24, 1982.

Stevens A, Lowe J: *Histology.* New York, Gower Medical, 1992.

Tavassoli M, Minguell JJ: Homing of hemopoietic progenitor cells to the marrow. *Proc Soc Exp Biol Med* 196:367–373, 1991.